Nuclear Weapons in the Information Age

Nuclear Weapons in the Information Age

STEPHEN J. CIMBALA

continuum

Continuum International Publishing Group

The Tower Building	80 Maiden Lane
11 York Road	Suite 704
London	New York
SE1 7NX	NY 10038

www.continuumbooks.com

ISBN: 978-1-4411-2684-9 (PB)

978-1-4411-8197-8 (HB)

Library of Congress Cataloging-in-Publication Data

Cimbala, Stephen J.

Nuclear weapons in the information age/Stephen J. Cimbala.

p. cm.

Includes bibliographical references and index.

ISBN-13: 978-1-4411-8197-8 (hardcover: alk. paper)

ISBN-10: 1-4411-8197-0 (hardcover: alk. paper)

ISBN-13: 978-1-4411-2684-9 (pbk.: alk. paper)

ISBN-10: 1-4411-2684-8 (pbk.: alk. paper) 1. Nuclear crisis control. 2. Information warfare. 3. Nuclear arms control. 4. Nuclear nonproliferation. I. Title.

JZ5665.C56 2012

327.1'747–dc23

2011034353

Typeset by Deanta Global Publishing Services, Chennai, India

Printed in the United States of America

Contents

Acknowledgment

The author gratefully acknowledges Penn State Brandywine campus for administrative support for this study. This book is dedicated to my wife Betsy, sons Chris and David, and daughter-in-law Kelly, with all my love.

Introduction

Nuclear weapons are the ultimate weapons of the industrial age – the age of mass destruction and total warfare. Postindustrial technologies, including those that support information based warfare, offer the prospect of prevailing in war at an acceptable cost by privileging smart, precision-guided weapons that minimize collateral damage. The coexistence of nuclear weapons with advanced conventional weapons and information-based concepts of warfare will be the most important military contradiction of the twenty-first century. The present study explains what this might mean by looking closely at six aspects or cases of this apparent contradiction or unplanned coexistence of two distinct arts of war, in the chapters that follow.

Twenty years beyond the end of the Cold War, nuclear weapons remain central to debates about the most significant threats to international peace and security. Evaluations of the importance of nuclear danger in this "second nuclear age" are not consistent as among policy makers and expert analysts. For example, U.S. President Barack Obama called in 2009 for the eventual abolition of nuclear weapons, joining a rising chorus of luminaries in government, academia, the clergy and other walks of life. Obama's national security and defense strategy included a large agenda of nuclear related objectives, including a strengthened Nuclear Non-Proliferation Treaty (NPT), U.S. ratification of the Comprehensive (nuclear) Test Ban Treaty (CTBT), sponsorship of an international agreement to cap the production of weapons grade material (the so-called Fissile Materials Cutoff Treaty, or FMCT). In addition, Obama moved quickly to negotiate with Russia the New START agreement on further reductions in the two states' strategic nuclear forces, signed by him and by Russian President Dmitri Medvedev in April, 2010, and subsequently ratified by the U.S. Senate in December, 2010.[1]

---

[1] For text of the New START treaty, see: *Treaty between the United States of America and the Russian Federation on Measures for the Further Reduction and Limitation of Strategic Offensive Arms* (Washington, D.C.: U.S. Department of State, April 8, 2010), http://www.state.gov/documents/organization/140035.pdf. For expert assessment of the Obama nuclear arms control agenda, especially the post-START and Comprehensive Test Ban Treaty ratifications, see Robert S. Norris, "The Senate and the START treaty," *Washington Times*, November 12, 2009, http://www.washingtontimes.com/news2009/nov/12/norris-the-senate-and-the-start-treaty/html.

Obama and other international leaders and nuclear security experts were calling, in summary form, for two kinds of restraints: on the growth in size of existing nuclear forces (vertical disarmament or arms limitation), and on the spread of nuclear weapons among countries that are technically capable and/ or politically interested in joining the ranks of nuclear weapons states (horizontal disarmament or arms limitation). However, some contended that Obama and others have exaggerated the degree of nuclear danger, in the past and present.

According to Professor John Mueller, nuclear weapons "have had a tremendous influence on the world's agonies and obsessions, inspiring desperate rhetoric, extravagant theorizing, and frenetic diplomatic posturing" although, in fact, "they have had very limited actual impact, at least since World War II."[2] Noting that military thinkers have struggled without success to define realistic battlefield uses for nukes, Mueller argues that ambitious plans for nuclear restraint like Obama's proposals are not necessary. For some time, nuclear weapons have been abolishing themselves, as states find their utility for deterrence and defense to be a wasting asset: "Nuclear weapons are already disappearing, and elaborate international plans like the one Obama is pushing aren't needed to make it happen."[3]

Kenneth N. Waltz sees an opposite trend in nuclear proliferation, compared to Mueller's forecast. Waltz regards nuclear weapons spread as unavoidable and the slow spread of nuclear weapons as preferable to no spread or rapid spread.[4] Basing his arguments on structural theory of international politics and on the history of states' behavior since 1945, Waltz argues that nuclear weapons make effective deterrents even in small numbers, provided forces are survivable against first strikes. According to Waltz, nuclear deterrence works precisely because it is based on uncertainty: "Uncertainty of response, not certainty, is required for deterrence because, if retaliation occurs, one risks losing so much."[5] This argument is related to Thomas C. Schelling's concept of the "threat that leaves something to chance": nuclear armed states have to be careful, not only to avoid a deliberate decision for nuclear first use or first strike, but also the likelihood of getting caught up in a process over which either, or both, would lose control.[6] According to Waltz, uncertainty-based nuclear deterrence has a strong track record:

---

[2] John Mueller, "Think Again: Nuclear Weapons," *Foreign Policy*, January/February 2010, http://www.foreignpolicy.com/articles/2010/01/04/think_again_nuclear_weapons

[3] Ibid.

[4] Kenneth N. Waltz, "More May Be Better," Ch. 1 in Scott D. Sagan and Kenneth N. Waltz, *The Spread of Nuclear Weapons: A Debate* (New York: W.W. Norton, 1995), pp. 1–45, esp. p. 42.

[5] Ibid., p. 24.

[6] See Thomas C. Schelling, *Arms and Influence* (New Haven, Ct.: Yale University Press, 1966), esp. p. 121, note 8, p. 105, note 4, and pp. 107–108.

It is now fashionable for political scientists to test hypotheses. Well, I have one. If a country has nuclear weapons, it will not be attacked militarily in ways that threaten its manifestly vital interests. That is 100 percent true, without exception, over a period of more than fifty years. Pretty impressive.[7]

Mueller's critique of establishment thinking about nuclear weapons offers a cautionary tale against exaggeration of nuclear danger, but not necessarily a viable policy or strategy story for heads of state and defense ministers. Policy makers and their military advisors and commanders are required to plan against a range of possibilities, including some outcomes of low probability but, nevertheless, high cost. The attacks of 9/11, for example, concentrated the minds of policy makers, not only about the capability of terrorists for making strategic attacks against the American homeland, but also about the potential for even greater destruction if nonstate actors acquired nuclear weapons. Even terrorists armed with nuclear weapons, however, could not threaten the entirety of civilization in the Northern Hemisphere, as the nuclear armed Americans and Soviets were thought to have done during the High Cold War.

In contrast to the argument that present and past nuclear dangers have been exaggerated, Michael Krepon contends that, in the second nuclear age, the character but not the nature of nuclear danger has changed. Nuclear danger now lies, not in the possible outbreak of global nuclear war between the United States and the former Soviet Union, but in the bracket creep of interest in, and possible acquisition of, nuclear weapons by states and nonstate actors with antisystemic goals and radical political agendas. As he notes, encouraging signs and negative trends with respect to nuclear risk reduction appear simultaneously:

> The opportunities for the most dangerous weapons and materials falling into new hands remain great, and yet there have been no mushroom clouds or acts of nuclear terrorism – yet. Although good fortune could end quickly and repeatedly, it remains difficult for states and extremist groups to succeed in the multiple steps required to obtain the bomb.[8]

Notwithstanding the technical and political barriers to nuclear acquisition, Krepon identifies nine "negative drivers" or possibly game-changing developments that could, in combination or separately, increase the degree of nuclear danger. These negative drivers include, for example: (1), the next

---

[7] Waltz, quoted in Amitai Etzioni, "Can a Nuclear-Armed Iran Be Deterred?," *Military Review*, May–June 2010, pp. 117–125, citation p. 120.

[8] Michael Krepon, *Better Safe than Sorry: The Ironies of Living with the Bomb* (Stanford, Calif.: Stanford University Press, 2009, p. 137).

use after Nagasaki of a nuclear weapon in interstate warfare; (2), failure to reverse North Korea's nuclear weapons program or to contain Iran's apparent drive for nuclear weapons; (3), the breakdown of civil authority in Pakistan together with a radical change of government; (4), an act of nuclear terrorism by extremists against a nation state.[9] Against these and other potentially disruptive forces, Krepon offers the hope of experience gained during the Cold War. U.S. policy makers and military leaders got through the Cold War to a successful outcome, including the peaceful deconstruction of the Soviet Union, by employing a combination of carrots and sticks: deterrence; military strength, containment, diplomatic engagement, and nuclear arms control. With appropriate adjustment for the challenges of the second nuclear age, compared to the first, the international community can use these same five sets of tools to contain and ultimately reduce the degree of global nuclear danger.[10]

Krepon's nuanced approached to the diagnosis of nuclear danger and its possible remedies gives important recognition to the complexity of the U.S. policy making process in defense and national security policy. States can be expected to improve, but not necessarily to transcend, their habitual modes of policy making and their national characters in relating leaders to followers. A Vladimir Putin would be as inconceivable as a U.S. head of state as would Nancy Pelosi be as leader of the majority party in the Russian Duma. The U.S. policy-making process, founded upon a Constitution that purposely divided power and authority among separate branches of government and between state and federal authorities, puts deliberate obstacles in the path of large change. Unless driven by a sense of emergency and a tailwind of popular support, the U.S. Congress and Executive Branch exhibit inertial forces sufficient to defy gravity.

Obama's ambitious agenda for nuclear limitation was consistent with Professor Graham Allison's statement of the "three nos" that U.S. arms control and disarmament policy would have to achieve to prevent nuclear terrorism. According to Allison, the "three nos" are: no loose nukes; no new nascent nukes; and, third, no new nuclear weapons states.[11] The immediate challenges are provided by North Korea and Iran, but other candidate nuclear weapons states might arise within the next decade or so, depending on the perceived balance of benefits and risks in doing so. Nuclear weapons can appeal to states for many reasons: for deterrence of

---

[9] Ibid.

[10] Ibid., pp. 174–212.

[11] Graham Allison, *Nuclear Terrorism: The Ultimate Preventable Catastrophe* (New York: Times Books – Henry Holt, 2004), p. 141 and passim.

hostile states, including possibly nuclear powers; for prestige and a seat at the head table among international powers; for domestic politics, including the rallying of national pride in scientific or military accomplishment; and, not the least, for "access denial" or deterrence against powers superior in conventional forces who might otherwise intervene in a state's regional backyard – or, worse, in its internal politics. Russia might be an example of the first kind of access denial use for nuclear weapons, and Pakistan, an example of the second kind.

The United States might heed the axiom "be careful what you wish for" in pursuing arms control and disarmament objectives. As the world's singular military superpower in conventional warfare, the United States might appear to be advantaged by an international regime in which nuclear weapons were marginalized or eliminated. However, the objective of deterrence, the avoidance of war by means of armed persuasion, is not necessarily advanced by trading in nuclear for conventional forces. And some conventional wars are better deterred than fought even for the "victorious" side. Nuclear weapons are instruments of influence by virtue of their effective nonuse. For any state contemplating an attack on another nuclear armed state, the defender's nuclear weapons push back against nuclear blackmail and, if mated to survivable delivery systems, provide for a response in kind. Since so few nuclear weapons can do so much societal damage, opposed general staffs and policy makers are denied the opportunity to write a script for victory at an acceptable cost.

The problem of victory denial at an acceptable cost is one that affects not only a state's estimates of its own predicament, but also its commitment toward any allies dependent upon its protection. The first nuclear age was characterized by global U.S.–Soviet conflict, with states lining up under the protective nuclear umbrella of one side or the other. The second nuclear age is obviously more fluid, with U.S. extended nuclear deterrence now stretched to include 27 allies in NATO and other Middle Eastern and Asian countries. So, for example, on the topic of conventional wars that are better deterred than actually fought, a war in Korea nominates itself as a strong candidate. Were North Korea to engage the armed forces of South Korea, supported by the United States, without nuclear weapons, the latter would eventually prevail, but at terrible cost in South Korean lives and, quite possibly, the defenestration of the Kim family regime in North Korea. Even the self-destruction of the North Korean state would present boggling problems for South Korea, China and the United States, including the possibility of a shared responsibility for postconflict reconstruction of a failed North Korean state with Russia looming over the horizon like a brooding omnipresence.

So, in the case of North Korea, we want to achieve denuclearization of the Korean peninsula without war: even though North Korea might "lose" the conventional war and, in the process, its nuclear weapons and infrastructure.[12] It might make sense for the United States and other powers to engage North Korea by offering an official war termination for the Korean war, together with additional economic and political incentives, in return for Pyongyang's disarmament and dismantlement. North Korea is politically isolated, and this provides its nuclear interlocutors with leverage, especially China, if they choose to use it. Diplomatic engagement with North Korea on denuclearization requires infinite patience on the part of the United States and its negotiating partners (Russia, China, Japan and South Korea), and the alternating bursts of cooperation and contumacy on the part of North Korea can send diplomats running to their psychiatrists. However, it's clear enough that, regardless of domestic upheavals, Pyongyang wants a guarantee against U.S.-instigated regime change (a la Iraq) as a preface to any concessions on its part.[13]

Iran is another case where the United States and allied European powers, including Russia, would prefer to resolve the issue of Iran's march toward the acquisition and deployment of nuclear weapons without having to engage in conventional warfare.[14] The situation with respect to U.S. extended deterrence against Iran, versus that with regard to North Korea, is more nuanced. South Korea is not a nuclear weapons state, but Israel and other possible targets for Iranian nuclear weapons, including American NATO allies, possess their own nuclear arsenals. The U.S. nuclear umbrella might be necessary to deter an Iranian nuclear threat against, or attack on, a nonnuclear ally: e.g., Iraq, or Saudi Arabia.

On the other hand, Iran is as likely as other nuclear weapons states have been to appreciate that nuclear forces are better for deterrence than for actual use: especially for access denial of United States or other military intervention against Iran's interests in the Persian Gulf. Behind a revetment of nuclear deterrence, Iran could use the threat or actuality of conventional war against regional rivals in order to assert its regional hegemony. The opportunity for Iranian regional assertiveness presents itself, ironically, because two of Iran's

---

[12] Expert assessment appears in Andrew Scobell and John M. Sanford, *North Korea's Military Threat: Pyongyang's Conventional Forces, Weapons of Mass Destruction, and Ballistic Missiles* (Carlisle, Pa.: U.S. Army War College, Strategic Studies Institute, April 2007).

[13] See Leon V. Sigal, "Let's Make a Deal," *The American Interest Magazine*, January–February 2010, http://the-american-interest.com/article-bd.cfm?piece=767, for additional perspective and background on this topic.

[14] For a discussion of military options in dealing with Iran, see Etzioni, "Can a Nuclear-Armed Iran Be Deterred." For possible scenarios if diplomacy fails, see Abdullah Toukan and Anthony Cordesman, *Iran, Israel and the Effects of a Nuclear Conflict in the Middle East* (Washington, D.C.: Center for Strategic and International Studies, June1, 2009).

worst enemies, the Taliban and Saddam Hussein, have been removed from power in Afghanistan and Iraq.

Nuclear weapons in the hands of leaders in Tehran or in Pyongyang, and perhaps spreading wider throughout both regions, invite some favorable glances on the part of policy makers and military planners toward missile defenses. Defenses against ballistic missiles emerged from the Cold War on a low burner. The high profile advocacy for nationwide antimissile defenses that characterized the Reagan defense program gradually gave way to more limited aspirations for systems that would deflect light attacks or accidental launches. Both the Bush I and Clinton administrations kept the flame of research and development for ballistic missile defense (BMD) burning, but neither showed great enthusiasm for imminent deployment. Congress passed legislation in 1998 calling on the President to deploy missile defenses at the first practicable opportunity, but Clinton departed office without making a decision and passed the baton to Bush II.

George W. Bush, in contrast to the first President Bush and Clinton, declared himself in favor or prompt deployment of national and global missile defenses in stages, beginning in 2004. He also removed the United States from commitment to the ABM Treaty of 1972, regarded by nuclear arms controllers as the touchstone of Cold War arms control but as obsolete and irrelevant by Bush II and his advisors. Russia was initially reticent about its disagreement with the United States over the decisions to depart the ABM Treaty and begin deploying missile defenses, but Russian President Putin went ballistic in 2007 when the United States announced plans to deploy components of its global missile defense system in Europe – specifically, in Poland and in the Czech Republic. U.S.–Russian disagreements over this proposal and other issues poisoned relations until the end of the Bush II administration. President Obama, as part of his "reset" of relations with Russia, revamped the U.S. missile defense plan for Europe, to emphasize sea based antimissile interceptors that presumably posed less of a threat to Russia's nuclear retaliatory force than the original Bush plan, with ground based interceptors deployed in Poland and Romania.[15]

However, the Obama missile defense shift may have simply postponed, instead of having resolved, conflict between Russia and the United States (and NATO) over missile defenses. Follow-on stages of the proposed European BMD program include increasingly capable Standard missiles deployed on land as well as at sea. An advanced version of the SM-3, under development since 2006, will apparently have the range to defend all of NATO from only a few

---

[15] Michael D. Shear and Ann Scott Tyson, "Obama Shifts Focus of Missile Shield," *Washington Post*, September 18, 2009, www.washingtonpost.com, downloaded October 22, 2009.

sites. Launched from land or at sea, SM-3s would provide flexibility in locating the launch point optimally, between the threat launch and the defended area – enabling the intercept of attacking missiles early in flight. In addition, the U.S. Missile Defense Agency's Phased Adaptive Approach to deployment offers to allies (including Poland and the Czech Republic) the possibility of collaboration on missile defenses, including the hosting of sites or the providing of BMD-supportive infrastructure. Russia might even get into the act, if radars at Gabala (Azerbaijan) and Armavir (Russia) were linked to the BMD sensor network as part of a comprehensive warning system pointed southward.[16]

Russia's response to U.S. missile defenses deployed in Europe or elsewhere will depend on the overall climate of U.S.–Russian relations more broadly construed. Russian threat perceptions are a movable feast.[17] Central to Russian assessment of U.S. and NATO intentions are Russia's fundamental interest in former Soviet security space and, as well, its desire to be accepted by the United States and NATO as an equal partner in drawing the present and future maps of European and Central Eurasian security (and, in fact, Russia might prefer to be a dominant partner in Central Eurasia).[18] However, Russia lacks the conventional military assets to support a geostrategic equivalence with NATO in jockeying for position in future European security space. Therefore, Russia must rely on its nuclear forces, including its transoceanic missiles and bombers, for the appearance of military-strategic parity with NATO. In addition, Russian military doctrine, unlike Cold War Soviet doctrine, no longer forswears the option of nuclear first use in cases of conventional conflicts that pose a vital threat to Russian national interests.[19]

Arguably, then, missile defenses will play with uncertainty into U.S.–Russian and broader discussions about nuclear arms reductions, disarmament and nonproliferation. Technologies for tactical or theater missile defenses are more advanced than those for defense against missiles of intercontinental range. NATO and Russia might find room for cooperation in warning and intelligence sharing with respect to out-of-the-area nuclear threats, including a possible

---

[16] See *Unclassified Statement* of Lieutenant General Patrick J. O'Reilly, USA, Director, Missile Defense Agency, Before the House Armed Services Committee (Washington, D.C.: U.S. House of Representatives, House Armed Services Committee, October 1, 2009).

[17] Possibilities are discussed in Stephen J. Blank, *Russia and Arms Control: Are There Opportunities for the Obama Administration?* (Carlisle, Pa.: Strategic Studies Institute, U.S. Army War College, March 2009).

[18] Pertinent insights and judicious assessment appear in Olga Oliker, Keith Crane, Lowell H. Schwartz, and Catherine Yusupov, *Russian Foreign Policy: Sources and Implications* (Santa Monica, Calif.: RAND Corporation, 2009).

[19] For analysis of recent developments, see "Russia's Message on Reshaping Its Nuclear Doctrine," *Stratfor.com*, October 15, 2009, in *Johnson's Russia List* 2009 - #190, October 15, 2009, davidjohnson@starpower.net.

one from Iran. As a candidate member of the four horsemen of nuclear policy analysis, BMD is more of a technological promissory note than a strategic breakthrough – at present. Even improved technologies for missile defense, unless based on new physical principles and operating at the speed of light, will not change the deadly arithmetic that it is easier to launch devastating nuclear attacks than it is to blunt those attacks in progress. Therefore, no silver bullet can repeal the nuclear revolution: only a combination of international diplomatic collaboration, political resolution, and military-strategic rationality can continue to reduce the risks of nuclear war in the second nuclear age.

II Plan of the book

Nuclear regimes are patterns of institutions, behaviors and values that give structure and coherence to the international and domestic policy-making processes pertinent to nuclear weapons. These alternative regimes are not equivalent to the products of political or military forecasting. No formula exists by which we can extrapolate with scientific certainty from the nuclear present to the nuclear future. As Chapter 1 explains, however, there is a nuclear past, and from that experience we can draw some lines around those nuclear regimes that are possible or probable, given present and foreseeable circumstances. Those regimes considered for further discussion in this chapter include: (1), a regime of mutual assured destruction (M.A.D.), or mutual deterrence; (2), a regime of nuclear primacy; (3), a regime of defense dominance; (4), a regime of nuclear abolition; and, (5), a regime of nuclear plenitude or abundance, along with tailored deterrence. Doubtless expert analysts and theorists can imagine other possibilities – admittedly, where the nexus between politics and nuclear weapons is concerned, there are few truisms that are uncontestable.

The coexistence of nuclear weapons with the tools of information warfare may be lethal under some conditions of crisis management, as explained in Chapter 2. Crisis management requires transparency and clear communication between potential adversaries, a willingness to consider the other side's point of view, clarification of potential misperceptions of the other side's intentions and capabilities, and an avoidance of time pressures leading to premature decision taking. Attacks on enemy computers and information systems during a crisis could compromise the decision making process of a nuclear armed opponent and lead to a mistaken decision for retaliation or a confused nuclear response by a hydra-headed nuclear command system after conflict had started. For these and other reasons, the continuing interest of the U.S., Russia, China and other major powers in offensive and defensive cyberwar (and related cyberdeterrence) takes place in a domain, cyberspace, with its own properties and rules of the road. Of immediate concern in Chapter 1, the "force multiplier" effect provided by cyberwar in conventional conflicts may have unintended and dysfunctional byproducts in nuclear crisis management.

Chapter 3 discusses the relationship between geography and nuclear arms control. The thesis of this chapter is that, regardless the political and military cast of nuclear activities, they take place inevitably within a geographical, and therefore geostrategic, context. The point might seem elementary, but awareness of geostrategy and the geographical context for nuclear deterrence, arms control and nonproliferation is often subordinated to other considerations. Geography is "just there" and taken for granted. But the very vocabulary for labeling nuclear delivery systems, including missiles and aircraft, is based on geography: that is, the estimated range or distance over which the plane or missile can deliver the nuclear warhead or warheads assigned to it. Thus we have intercontinental or transoceanic (customarily referred to as "strategic" in this context) delivery systems as well as those of intermediate, medium and short range (frequently identified as substrategic). A political aspect of geography is that it helps to define for a state its possible enemies and friends, and associated perceptions of threat or reassurance. Russia, for example, worries about the expansion of NATO since the end of the Cold War to some twenty-eight members and to the very borders of Russian territory. As another example, geography impacts directly upon the related issues of conventional and nuclear arms control in Europe, including the proposal by some NATO political leaders to remove all tactical or battlefield nuclear weapons deployed in Europe. Russia demands that NATO repatriate all U.S. tactical nukes currently deployed in Europe before Russia will consider denuking its own forces deployed in western Russia – the Kremlin argues that the United States is the only remaining state that currently deploys nuclear weapons outside its own national territory.

Chapter 4 takes up the problem of nuclear abolition. Nuclear abolition has been an expressed desire on the part of political activists, including scientists and prominent politicians, since the beginning of the nuclear age. Abolition has been given fresh momentum in the past two decades by the end of the Cold War, by the dissolution of the Soviet Union, and by the fast forward in information-based, high end conventional warfare, especially in the United States. Nuclear weapons, so tied to the Cold War politically and so unique in their capacity for prompt mass destruction, now seem to many influential former policy makers and military experts as atavistic and unnecessarily dangerous. In addition to the dangers of nuclear weapons spread among state actors, since 9/11 leaders are now attentive to the possibility that nonstate actors, including terrorists, may acquire and use nuclear or other weapons of mass destruction (WMD). The dangers of nuclear weapons spread among states and nuclear terrorism are connected. As more states, including those with regional gripes or hegemonic ambitions, acquire nukes, the possibility of illicit transfers of nuclear weapons technology and know-how to terrorists looms larger. Despite these apparent dangers, however, nuclear abolition has formidable political and military obstacles to overcome, including the mistrust of nuclear weapons states for one another's motives and sincerity as well as the difficulty of objective monitoring for a condition of nuclear "zero."

In Chapter 5, we review the implications of the New START agreement between the United States and Russia (ratified by the U.S. Senate in December, 2010). New START reduces allowable numbers of deployed U.S. and Russian strategic nuclear weapons and launchers below previously agreed levels and provides for a regime of inspection and verification. New START was negotiated within an entirely different political climate as between the United States and Russia, compared to the frosty Washington–Moscow relationship that dominated George W. Bush's second term. Therefore the significance of New START was more political than strictly military. It could, under propitious developments in politics and diplomatic engagement, open the door to further Russian–American strategic nuclear arms reductions and improved cooperation on nuclear nonproliferation. For Obama, New START was only part of a larger agenda for nuclear limitation that included U.S. adherence to the Comprehensive Test Ban Treaty (CTBT), reinvigoration of the Nuclear Non-Proliferation Treaty (NPT), and the creation of a global Fissile Materials Cutoff Treaty (FMCT) to prevent the use of fissile materials in weapons production. U.S. missile defenses, under Obama as under Bush, loomed as a swinging door for U.S.–Russian relations: a potential source of disagreement amid Russian fears of a U.S. technology that would negate Russia's nuclear deterrent; or, conversely, as another arena of U.S.–Russian cooperative security, fully compatible with other deterrence, nonproliferation and arms control initiatives.

In Chapter 6, the problem posed by North Korea as a fledgling nuclear weapons state is treated within the larger context of its implications for deterrence, nuclear arms control and nonproliferation. North Korea defies analysis according to conventional wisdom because the behavior of its regime is so unpredictable, and its motives, obscure. The decision making process in Pyongyang is enigmatic, even for close observers like South Korea and China. The greater danger is the eruption of a foreign policy crisis as between North and South Korea, combined with a fear of regime change on the part of North Korea. Regardless the nuances of the "Kim family regime," the management of nuclear restraint in Asia depends on nuclear containment, if not disarmament, of North Korea. There is no alternative to diplomatic engagement because the use of force against North Korea would have unacceptable consequences for

South Korea – even if North Korea's nuclear weapons are not brought into play. Favorable outcomes from diplomatic engagement with North Korea on nuclear issues, including possible rollback of its weapons and infrastructure, will require sustained commitment from China and Russia, especially the former. Success also requires the navigation of periodic crises between North and South Korea, inevitable speed bumps so long as the state of war between the two Koreas continues.[20]

Recognition of nuclear dangers in Asia does not imply that Europe is free of nuclear danger. As is noted in Chapter 7, continuing NATO deployments of "sub-strategic" U.S. nuclear weapons in Europe, and Russian substrategic nuclear weapons located in its western theaters of military operations, are part of a larger matrix of deterrence, defense and nuclear arms control issues. On one hand, leading politicians in Belgium and Germany have called for removal of an estimated 200 U.S. substrategic nuclear weapons, located in Belgium, Germany, Italy, the Netherlands and Turkey.[21] On the other hand, the Group of Experts panel on a new strategic concept for NATO noted in May, 2010 that, under current security conditions, "the retention of some U.S. forward-deployed systems on European soil reinforces the principle of extended nuclear deterrence and collective defence."[22] The issues of collective defense and extended deterrence involved both nuclear and conventional forces, as well as the question under what conditions NATO, or Russia, might engage in nuclear first use. Russia's 2010 Military Doctrine was less permissive, with regard to the stated conditions for nuclear first use, than Russian hawks had demanded or Western pessimists feared. But remaining weaknesses in Russia's conventional forces, unless remedied by ambitious military reform plans, invited excessive reliance by Russia on early resort to nuclear escalation and, therefore, resistance to substrategic denuclearization in Europe.

---

[20] An example is provided by the sinking of the South Korean warship Cheonan on March 26, 2010, in which 46 sailors were killed. A team of military and civilian investigators from Australia, Britain, South Korea, Sweden and the United States concluded that the South Korean ship had been sunk by a torpedo fired by a North Korean submarine. South Korea called for UN sanctions against North Korea and North Korea responded with predictable denials, threatening to respond to sanctions with "strong measures, including a full-scale war." See Choe Sang-Hun, "South Korea Publicly Blames the North for Ship's Sinking," *New York Times*, May 19, 2010, http://www.nytimes.com/2010/05/20/world/asia/20korea.html and David E. Sanger and Choe Sang-Hun, "North Korea Cuts All Ties With South," *New York Times*, May 25, 2010, http://www.nytimes.com/2010/05/26/world/asia/26korea.html

[21] Belgian and German calls for removal of U.S. tactical nukes are noted in "NATO Mission Statement Supports Retaining Tactical Nukes," *Global Security Newswire*, May 17, 2010, http://gsn.nti.org/gsn/nw_20100517_8029.php

[22] *NATO 2020: Assured Security; Dynamic Engagement*, Analysis and Recommendations of the Group of Experts on a New Strategic Concept for NATO (Brussels: NATO, May 17, 2010), p. 43.

Beyond New START, the possibility of additional success in U.S.–Russian nuclear arms reductions may require "out of the box" thinking compared to past practice. In Chapter 8, we discuss the possible conjunction of more drastic U.S.–Russian offensive force reductions and the deployment of missile defenses. Instead of proceeding incrementally from New START to the next stage of reductions in long range nuclear weapons, Russia and the United States might leap forward to a "minimum deterrence" regime based on a thousand or fewer operationally deployed weapons. A minimum deterrence regime could support political and military stability between Washington and Moscow at lower levels of armament than hitherto. In addition, the idea of minimum deterrence could also be proposed for adoption by most existing nuclear weapons states as the basis for a multilateral nuclear "concert" or constrained proliferation system.

This more ambitious proposal, of multilateralizing minimum deterrence, requires heavier political lifting among nuclear weapons states than prior experience would anticipate. But some states might support minimum deterrence as a less than utopian alternative to nuclear abolition that still leaves in the hands of powers the capacity for mass destruction, and therefore deterrence, many times over. Complicating the efforts of the United States and Russia to move toward minimum deterrence as the basis for offensive force reductions is the issue of missile defenses. U.S. plans for missile defense modernization and deployment, especially in Europe, have raised concerns in Russia that advanced missile defenses might nullify Russia's strategic nuclear deterrent. On the other hand, NATO has invited Russia to participate in discussions about the joint deployment of European antimissile defenses, presuming collaborative interest in deterring attacks from Iran or other states outside of Europe. The details of shared missile defenses development and deployment, as between NATO and Russia, involve both technical and political complexity. The likelihood of success in this endeavor also depends upon domestic politics within the various state capitals of NATO and in Moscow.

1

Alternative nuclear regimes

Introduction

In international politics, a regime is a collection of rules or behavioral expectations that provide a framework for the interactions among states or other actors. We can think of at least five alternative nuclear regimes that have characterized the past and present, or might conceivably evolve in the future. The term "regime" should not mislead. International regimes are dissimilar from the regimes that govern states. Regimes that govern states are presumed sovereign over a given territory, supported by a monopoly over the authoritative use of force within that territory. State regimes have as their constituents either citizens or other persons who have various degrees of legal claim on them, and obligations to them.

International regimes, including the nuclear regime shared by states, are intellectual constructs as much as they are physical and legal embodiments.[1] Nevertheless, international regimes are sometimes represented in institutional form: the Nuclear Non-Proliferation Treaty (NPT) is an example. Treaties are a frequent means of giving formality and concreteness to international regimes. Another means of concretizing international regimes is the creation of international organizations. Not all of these international organizations represent governments, although some do: the United Nations; NATO; the European Union; and so forth. Other organizations, termed NGOs (for

---

[1] For more on the concept of regimes as applied to nuclear weapons, see David W. Tarr, *Nuclear Deterrence and International Security: Alternative Nuclear Regimes* (White Plains, N.Y.: Longmans, 1991). Also useful and more contemporary is Michael Krepon's discussion of alternative nuclear futures in Krepon, *Better Safe Than Sorry: The Ironies of Living with the Bomb* (Stanford, Calif.: Stanford University Press, 2009), pp. 156–173.

nongovernmental organizations), also contribute to the development of norms and expectations about international behavior. For example, transnational NGOs such as Doctors without Borders and the Red Cross have contributed to the establishment of conventions about international humanitarian relief and medical treatment.

Nuclear regimes can be classified as one or more of the following: (1) a regime of mutual assured destruction (M.A.D.), or mutual deterrence; (2) a regime of nuclear primacy; (3) a regime of defense dominance; (4) a regime of nuclear abolition; and, (5) a regime of nuclear plenitude or abundance, along with tailored deterrence.

Mutual deterrence

The first regime, of deterrence based on mutual assured destruction, characterized the Cold War from the time that both the United States and the Soviet Union acquired a survivable nuclear second strike capability. Mutual assured destruction is thus based on assured retaliation.[2] The prospective attacker or first striker is made aware that, regardless of his choice of attack strategy or the extent of the defender's surprise, the attacker will receive an unacceptable level of damage in retaliation. In the case of the mature Cold War American and Soviet intercontinental range or "strategic" nuclear forces, a nuclear war begun by either side would have resulted in prompt and delayed retaliation that inflicted historically unprecedented damage. Because both states' leaders knew this and could find no recipe for "victory," or prevailing in nuclear combat at an acceptable cost, they settled for stalemate.[3]

The regime of mutual assured destruction or assured retaliation has not really been tested apart from the nuclear bipolar conditions of the Cold War. The United States and the Soviet Union dominated the international

---

[2] According to Lawrence Freedman, the term "assured destruction" first appeared in U.S. defense policy discussions in 1964; the concept was originally described as "assured retaliation" but the latter was thought to be too bland. Mutual assured destruction was essentially the same concept as the idea of a "stable balance of terror" from the late 1950s. See Freedman, *The Evolution of Nuclear Strategy* (New York: St. Martin's Press, 1983), pp. 246–248.

[3] Robert Jervis, *The Meaning of the Nuclear Revolution: Statecraft and the Prospect of Armageddon* (Ithaca, N.Y.: Cornell University Press, 1989), pp. 74–106 explains the logic of mutual assured destruction. See also: Jervis, *The Illogic of American Nuclear Strategy* (Ithaca, N.Y.: Cornell University Press, 1984), esp. pp. 47–63. For assessments of post-Cold War nuclear deterrence, see: Krepon, *Better Safe Than Sorry*, esp. pp. 94–132; Patrick M. Morgan, *Deterrence Now* (Cambridge: Cambridge University Press, 2003); Colin S. Gray, *The Second Nuclear Age* (Boulder, Colo.: Lynne Rienner, 1999); and Keith B. Payne, *Deterrence in the Second Nuclear Age* (Lexington, Kentucky: University Press of Kentucky, 1996).

nuclear system and imposed, within their respective alliances, restraint on the nuclear aspirations of other states. The Cold War ended with five acknowledged nuclear weapons states (the United States, the Soviet Union (later Russia becomes the nuclear successor to the USSR), Britain, France and China, or the P-5 permanent members of the UN Security Council, and one de facto but unacknowledged nuclear weapons state – Israel. India and Pakistan were well on the way to acknowledged nuclear weaponization by the time of the Cold War endgame, and they went public with their status as nuclear weapons states with almost simultaneous tests in 1998. North Korea withdrew from international agreements restricting its nuclear development beginning in 2002 and conducted its first admitted nuclear weapons test in 2006.

Given the preceding résumé, one can argue that the amount of nuclear weapons spread among states, during the Cold War and even afterward, to the end of the George W. Bush administration, was thankfully limited. A total of nine either confessed or nonconformist nuclear weapons states existed at the end of the calendar year 2008. In addition, negotiations were in progress to reverse North Korean nuclear weapons capability by means of diplomatic engagement and economic incentives. Other good news, for adherents of peace based on assured retaliation, was the most important nuclear nonevent of the twentieth and twenty-first century (thus far). No nuclear weapon had been fired in anger since the U.S. attack on Nagasaki in August, 1945. This nonevent was a nuclear taboo that had lasted at least sixty years and counting.

Experts cautioned, however, that prior restraint was no guarantee of future forbearance. There were reasons to be skeptical that a regime of mutual deterrence built for the Cold War would work in a post-Cold War, and prospectively multipolar, nuclear world. First, the Cold War superpowers came to understand one another's nuclear policies and strategies as a result of experience, including trial and error (as in the Cuban missile crisis) and various rounds of nuclear arms control negotiations. As between the Cold War Americans and Soviets, and to some extent carried forward between post-Cold War Americans and Russians, there grew up a dialogue of shared expectations as to what was worth arguing about – and in what language. After Soviet premier Nikita S. Khrushchev, neither United States nor Soviet or Russian leaders resorted to unvarnished nuclear blackmail of the kind that the ebullient Khrushchev paraded before an international audience. It may overstate to describe the Cold War as a long peace, since a number of hot wars were fought under its aegis – but they were not fought between Americans and Russians, they were not fought in Europe, and they were not fought with nuclear weapons.

In other words: deterrence seemed to have "worked," but its success was possibly politically overdetermined by a number of factors. These included: strategic nuclear bipolarity; mutual learning, although not necessarily full agreement, about the practice of deterrence and the management of nuclear operations; a supportive political and legal framework, including the various arms limitation and reduction talks, international agreements on nonproliferation, nuclear testing and other matters; and a growing familiarity on the part of policymakers in Washington and in Moscow; and, most important for the prevention of nuclear weapons spread, the willingness of other states to accept the nuclear defense guaranty of the United States or the Soviet Union instead of pursuing their own independent nuclear forces.

As the reader will have guessed, the incentive structure for nuclear restraint just described has been whittled away by the end of the Cold War and the demise of the Soviet Union, in addition to other post-Cold War developments. New or aspiring nuclear powers lack the restraints of nuclear bipolarity, the experience in nuclear policy and strategy making, and the time for shared nuclear learning among potential adversaries that was provided the Cold War Americans and Soviets. In addition, there is the phenomenological problem about deterrence that remains unresolved among theorists and policy makers. How do we know reliably when deterrence has worked, as intended, instead of having succeeded fortuitously or proved to be irrelevant in the exigent circumstances? The absence of war or threat of war does not necessarily prove the case for the "success" of deterrence. War might not have been intended, or if intended, incentives apart from deterrence might have mattered more. Proving the chain of cause and effect for a "non-event" often exceeds the capacity of much social science and historical research. The absence of war, in short, does not prove the presence of deterrence.

Deterrence can apply to wars that might have been fought with conventional as well as nuclear weapons. The track record of deterrence with respect to conventional wars is uneven. Deterrence based on the certainty of unacceptable retaliation characterizes the preferred practice for nuclear weapons. On the other hand, conventional deterrence depends on something altogether different: the credible threat posed by the defender to prevail in battle against the attacker. History is littered with cases in which apparently weaker states were undeterred from attacking presumably stronger military opponents, and in many of these cases, the attacker lost. In both World Wars of the twentieth century, political leaders and their military advisors gambled on prompt offensives based on battles of annihilation against formidable opponents or coalitions (for example, Germany in World War I against France, and Germany, again, against the Soviet Union in World War II). In addition, historians of both wars have documented major deficiencies in the intelligence collection and estimation on the part of great powers preceding both world wars. In both conflicts, some states plunged into fateful battles and wars that destroyed their regimes and empires with scant awareness of their enemies' art of war, potential for fighting a protracted war, or military–strategic doctrines.

Thus far, nuclear deterrence has fared better than conventional deterrence based on experience, but the nuclear deterrence has had a much shorter shelf life. The critical experiment for the efficacy of nuclear deterrence based on assured retaliation has yet to be performed. That experiment would include one, or more, of the following attributes: (1), nuclear armed coalitions, in addition to individual state actors, on the playing field of immediate deterrence; (2), defaults in reciprocal communication between or among crisis time adversaries; (3), political leaders not well informed about their adversaries' true intentions or military capabilities; (4), political leaders not well informed about their *own* military capabilities, and limitations; (5), military leaders who are intermittently deferential to heads of state and government, and others who might constitute a self-referential professional group on military, including nuclear, matters; (6), political leaders motivated by compromise-resistant nationalistic, religious or other motives for war that create cognitive blinders to options for peace; (7), bureaucracies unable to process information and decisions in good time, relative to the pace of an unfolding crisis and the inertia of movements toward war; (8), mistaken beliefs in windows of opportunity for success based on offensive strategies and preemption; (9), prewar misperception and miscommunication, based on poor diplomacy, hidden agendas on the part of leaders, flawed intelligence or other factors; (10), absence of any plan with branches or plan "B" in case the plan designed for the initial period of war proves faulty, or the war becomes more protracted than anticipated; and (11), an inadequate survey of options short-of-war, prior to the onset of a crisis, and a decision making process that falls short of adaptive planning and learning requirements under duress. This short list of possible flies in the ointment of nuclear deterrence is neither imaginary nor improbable: remove the adjective "nuclear" and we have seen it before, in the crisis management behavior of states committing conventional deterrence malpractice.

On the other hand, establishing that conventional deterrence has somehow "failed," within the larger causal framework of war and peace, is perhaps a self-referential exercise for deterrence theorists. Deterrence has omnivorous conceptual reach: it has been used to describe behaviors from parenting to the avoidance of nuclear Armageddon. Even the idea of deterrence as applied to military affairs is sometimes a gap filler for want of better ideas. The empirical evidence to support hypotheses of "success" in deterrence, conventional as well as nuclear, is almost always contestable. One problem is the nature of the dependent variable. It appears to be a simple binary choice: war or no war. But the appearance of simplicity is deceptive. Between the final decision for or against war and the chain of cause and effect leading up to war are many intervening variables. Some of these intervening variables have to do with technology, others with politics, and others again with psychology or philosophy. Leaders' or theorists' claims that war broke out because deterrence failed have a tautological quality: one could with equal justice argue that deterrence failed because war broke out.

In addition, deterrence evolved as a concept applied to nuclear strategy as the Cold War progressed. From a concept designed as a way station to a world of more compelling and empirically verifiable generalizations, deterrence became a lingua franca among military strategists. In some cases deterrence overlapped with policy prescriptions so much that the two could not be distinguished without a magnifying glass. U.S. nuclear deterrence theorists even claimed at various times that Soviet military theory coincided with their own American understandings of the concept, despite considerable evidence to the contrary in Soviet military writing. From a bridge between policy analysis and applied science, deterrence became an all purpose solvent for military and strategic problems. In some ways this was odd, since deterrence was and is fundamentally a psychological concept: and psychologists, for the most part, disdained nuclear weapons as an abomination.

Some military thinkers and weapons scientists have embraced nuclear weapons, and the idea of peace by means of nuclear deterrence, with reluctance since the dawn of the atomic age. Other experts and concerned lay people have called for alternatives to dependency on nuclear threats for survival. The regime of nuclear deterrence by means of assured retaliation is therefore not holding its ground uncontested. But, proposed alternatives to assured retaliation or mutual deterrence must confront a large massif of inherited institutional, political and professional commitments in favor of nuclear weapons and deterrence on the part of current nuclear weapons states. Alternative nuclear regimes that have been proposed or feared as unwanted, but possible, outcomes include: (1), nuclear primacy; (2), defense dominance; (3), nuclear abolition; and, (4), nuclear plenitude, perhaps leading to nuclear anarchy.

Nuclear primacy

Nuclear primacy is the opposite of a condition of nuclear deterrence based on assured retaliation. In a condition of nuclear primacy, one state has acquired a nuclear first strike capability. The primary state can launch a first strike against any opponent with impunity. This usually means that the first striker can destroy enough of another state's nuclear retaliatory forces that the victim is denied an effective second strike. An effective second strike would inflict on the first striker damage that the attacker judged to be unacceptable by its standards. Since a few nuclear weapons can do a considerable amount of damage by historical standards, the requirements for nuclear primacy are high indeed. The retaliating state need only explode tens of weapons over several cities to provide for "unacceptable" damage against most putative attackers. By broad definition, the United States was in a condition of nuclear primacy from 1945 until 1949, when the Soviet Union detonated its first atomic device. Nuclear armed states can therefore attack nonnuclear weapons states with impunity if those defenders are not allied with another nuclear power.

Nuclear primacy appeared as an attractive option for some U.S. strategists in the early years of the Cold War. Throughout the Cold War, both American and Soviet strategists and politicians worried about the potential vulnerability of their strategic nuclear forces to surprise attack. To increase force survivability, the Soviet Union and the United States deployed nuclear charges in different basing modes and across three kinds of intercontinental delivery systems: land based ballistic missiles (intercontinental ballistic missiles), or ICBMs; submarine launched ballistic missiles, or SLBMs; and bomber delivered weapons, including gravity bombs, short range attack missiles (SRAMs), and air-launched cruise missiles (ALCMs). The assumption was that diverse basing modes, together with some redundancy in numbers relative to the attacker's prospective options, would provide for a reliable second strike capability under all foreseeable circumstances. Of course, it became necessary to modernize each kind of delivery system over the decades, and competition within and among the military services of the U.S. and Soviet armed forces, for funding and for missions, was a characteristic of domestic politics in Washington and in Moscow.

The Cuban missile crisis was arguably the first true "road test" for nuclear primacy. The United States faced down the Soviets in October, 1962 in a tense confrontation that left its mark on future U.S.–Soviet relations until the end of the Cold War. At this time, the United States had an approximate advantage of sixteen to one over the Soviet Union in deliverable nuclear warheads. If nuclear war were algebra, the United States could have "won" a nuclear conflict against the Soviet Union in 1962: the United States would have absorbed far less damage relative to the destruction that it could have inflicted on the Soviet Union. However, this was not how President Kennedy and his advisors in the ExComm crisis advisory group reckoned their options.

Kennedy's major concern was not to prevail in a nuclear war against the Soviet Union, but to remove the Soviet missiles from Cuba without war. If the political crisis had escaped his and Soviet Premier Khrushchev's control,

U.S. air strikes or an invasion of Cuba might have been followed by Soviet countermoves in Berlin, leading to a NATO–Soviet confrontation and possibly to a third world war. The reality of this stark possibility of mass destruction, either on European or American soil, caused Kennedy and Defense Secretary Robert McNamara to employ coercive diplomacy to remove the Soviet missiles instead of more dangerous alternatives. The choice was fortuitous.

We now know, as a result of post-Cuban crisis research over several decades, that President Kennedy and his advisors had only partial information about Soviet nuclear weapons deployed in Cuba. In addition to the warheads deployed for possible use with Soviet intermediate and medium range ballistic missiles on that island, Moscow had also authorized the deployment of tactical nuclear weapons available for possible use by Soviet ground forces commanders in Cuba. Under some circumstances, these local Soviet commanders would have had authority to use these short ranges, nuclear capable missiles without specific authority from Moscow.

In addition to incomplete intelligence on both sides about one another's capabilities and intentions, the Cuban missile crisis was also marked by unexpected glitches in command and control. A U-2 reconnaissance aircraft "strayed" into Soviet air space at one of the tensest moments of the crisis, causing Soviet air defenses to scramble and Soviet leaders to wonder whether this was a precursor reconnaissance to an Americana nuclear first strike. The official explanation from United States sources was that the U-2 wandered off course while sampling readings from the Northern Lights. Another misstep occurred when Soviet air defense commanders in Cuba authorized the shooting down of a U-2, causing the death of its pilot and leading President Kennedy to wonder whether this was an intentional escalation by Moscow (it was not). In a third anomalous event during the crisis, a U.S. ICBM test firing from California into the Pacific was allowed to proceed on schedule, despite the possibility that observers might attribute to the test a much more sinister meaning, in the exigent circumstances.

In addition to the above, what is even more interesting from the standpoint of aspirations for nuclear "primacy" was the reaction of Kennedy and his advisors to the available nuclear war plans. The nuclear war plan passed from Eisenhower to Kennedy apparently called for massive attacks on the Soviet Union and other states in Europe, plus China (then assumed to be in lock step with Soviet plans for world domination). Casualties from execution of this plan would have been in the many hundreds of millions. Horrified at this overkill, Kennedy and McNamara were in the process of redefining the nuclear war plan to include a variety of options involving more selective attacks. McNamara even flirted with the idea of restricting a war with the Soviet Union to "counterforce" exchanges with both sides forbearing from follow-up attacks on cities. The Cuban missile crisis erupted before the administration was able to fully implement its thinking about limited nuclear options and damage limitation within war. What was on the shelf, in terms of nuclear options actually available to the United States in October, 1962, was insufficiently nuanced for Kennedy's needs and actually repugnant to Secretary McNamara. The existing SIOP (Single Integrated Operational Plan for nuclear war) was declared a nonstarter for crisis management purposes – despite U.S. nuclear "superiority" in numbers and survivability of weapons.

The Cuban missile crisis helped to sober the Cold War superpowers into movement toward a regime of mutual deterrence and away from claims or aspirations for nuclear primacy. The end of the Cold War and demise of the Soviet Union would seem to have added credibility to the regime of deterrence and cast additional doubt on the regime of nuclear primacy. But nuclear primacy has not gone away. It has reappeared in a post-Cold War hypothesis that the United States might have a future first strike capability against post-Soviet Russia. In this hypothesis, advanced by U.S. analysts and provocative of a great deal of commentary in Russia, a future Russia might be unable to maintain its nuclear deterrent credibility for two reasons: (1), insufficient funds to provide for both the numbers and qualities of survivable weapons and launchers; and, (2), a U.S. technological lead that would sideswipe Russia's deterrent, including possibly information-enhanced, global and precision strike systems with conventional and-or nuclear weapons.

Theoretical models of U.S. nuclear primacy can be constructed, but how realistic are they?[4] Russia would have to be virtually broke and strategically inept to permit its nuclear deterrent to lapse into such a vulnerable condition – even in the abstract world of calculations. In reality, a U.S. nuclear first strike against Russia has a variety of political and military impediments to overcome: (1), it would be inconsistent with American values and with American experience since the end of World War II; (2), military options other than a nuclear first strike could accomplish almost any plausible U.S. policy aim or military objective, short of mass destruction for its own sake; (3), Europe, especially U.S.-allied Europe in the form of NATO, figures into the equation. Russia, even if denied its nuclear second strike capability against North America and other targets requiring intercontinental launchers, could inflict devastation beyond historical precedent on European cities with its remaining theater and tactical nuclear weapons. Strikes against Russia by British and French nuclear forces would surely follow. What was left of Europe's economy

---

[4] See, for example, Keir A. Lieber and Daryl G. Press, "The Rise of U.S. Nuclear Primacy," *Foreign Affairs*, March/April 2006, http://www.foreignaffairs.org/20060301faessay85204/keir-a-lieber-daryl-g-press/html.

would be a basket case, inflicting trans-Atlantic pain on the U.S. economy as well; (4), a U.S. president who ordered a nuclear first strike against Russian territory, without first trying diplomacy or other-than-nuclear military options, would encounter resistance from his or her military advisors, and therefore, from Executive Branch and Congressional doubters; (5), despite their high levels of professionalism and patriotism unquestioned, the military operators of American nuclear forces are human beings with consciences and humane values. Some might resist orders for preemptive nuclear war, others could engage in bureaucratic stalling tactics, and others would leak to the press presidential intentions that (in their view) bordered on madness.

If one doubts this, consider the evident resistance of American military lawyers to the George W. Bush administration policies on aggressive interrogations and preventive detentions for suspect terrorists. In addition, The American military is apolitical, but it is not demonstrably suicidal: nuclear force commanders might consider that the "fog of war" and Clausewitzian "friction" would result in a gap between preemptive optimism and exigent results. The combination of moral doubts, legal ambiguities, military conservatism and collateral damage inherent in nuclear attacks could even lead to the impeachment of a President who was contemplating a nuclear first strike. Granted, the Bush II administration nuclear policy guidance did hold open the option for conventional or nuclear preemption, against terrorists equipped with weapons of mass destruction and capable of attacking the American homeland – as well as against states that supported and succored such terrorists. That bit of fallout from 9/11 is a form of preemptive self-defense that is arguably consistent with international law. A massive nuclear first strike against Russia would not be either morally, legally or strategically equivalent to preemptive self-defense against nuclear armed terrorists.

Another problem with aspirations for global nuclear primacy is that a state attempting to achieve these heights would almost certainly be resisted by competitors. Failure by others to resist ambitions, on the part of a single state or center of authority, for nuclear primacy, would be to cede sovereignty over their respective foreign policies. The singular authority in possession of nuclear weapons could enforce its writ over any issue or opposed regime. While some states might bandwagon with this nuclear sovereignty, other states would have to catch up or overtake the nuclear sovereign. Even the appearance of a state with military capability approximating nuclear primacy would allow that state a degree of coercive diplomacy and conclusive military threat that would not be available to others. Fears of this sort are among the reasons why post-Soviet Russia, in both good and bad financial times for military procurement, has insisted that nuclear force modernization remains a national defense priority.

Resistance to nuclear primacy does not necessarily come in the form of a nuclear arms race among the powers. Some states might decide to compete or checkmate the aspiring nuclear hegemony by means other than qualitative or quantitative nuclear competition. For example, states might try to leapfrog beyond the current technology for nuclear weapons and launchers. Advanced non-nuclear technologies for terrestrial, sea-based, airborne or space based missile intercept could be developed and deployed as an alternative to nuclear arms racing. Ironically the United States, which is the subject of contemporary fears about its offensive nuclear primacy, has also taken the lead in research, development and deployment of ballistic missile defenses. The George W. Bush administration announced in 2001 its intent to withdraw from the ABM Treaty, the cornerstone of Cold War arms control agreements. The U.S. Department of Defense began deploying components of its Global Missile Defense System in 2004. Being tested and deployed in pieces, the system is planned to include ground-based, sea-based, and airborne interceptors, as well as space-based components for reconnaissance and warning, communications, and command-control.

States such as Russia and China are not concerned about possible U.S. nuclear primacy only on account of America's offensive missile and bomber capabilities. They are also leery of the U.S. lead in missile defenses, defenses that, if good enough, might nullify the deterrent capability of China and Russia's nuclear retaliatory forces. The combination of superior offenses and defenses might provide the United States with a putative first strike capability that would be unattainable with offenses alone. There is more. A third aspect of U.S. defense modernization is the undoubted supremacy of American military technology for large scale conventional warfare, especially that fought over long distances and enabled by U.S. supremacy in C4ISTAR (command, control, communications, computers, intelligence, surveillance, targeting and reconnaissance) and precision strike.

It is the possible combination of U.S. superiority in offensive nuclear weapons, missile defenses and advanced technology conventional warfare that creates fears of American military primacy on the part of rivals and competitors. The point is often missed by those who search for the Rosetta stone of "nuclear" superiority over rivals, as did some Americans and Soviets during the high Cold War. If the Cold War proved anything, it demonstrated that nuclear superiority was a chimera. Nuclear powers found that their weapons were useful as deterrents against other nuclear powers that might use their own nuclear forces for coercion. But the diplomatic and military reach of nuclear forces, even for the overweight arsenals of the Americans and Soviets, was circumscribed by the realities of international politics.

Most hot wars that were fought during the Cold War were irregular conflicts in which at least one side was represented by insurgents or other nonstate actors. These wars were spawned by revolutions against colonialism, by ethno-nationalist or religious conflicts within states, or by other forces not amenable to nuclear deterrence. Exceptions included the "regular" wars in Korea from 1950–1953 and the U.S. and allied coalition war against Iraq in 1991. In neither of these two conflicts were nuclear weapons used or included in presidentially approved war planning. Admittedly, there were nuclear vapors in both cases. The Eisenhower administration is thought by some to have considered using nuclear weapons in Korea to break the military stalemate, and Eisenhower's pre-presidential campaign statement that "I shall go to Korea" allegedly betokened a willingness to consider nuclear first use against the DPRK or China. In fact, Eisenhower never seriously considered using nuclear weapons in Korea, essentially following the Truman policy of limiting the war to the Korean peninsula and to conventional weapons.

The nuclear vapors in the Gulf war of 1991 were the alleged tacit threats by U.S. Secretary of State James Baker and others to use nuclear weapons in response to any use by Saddam Hussein of chemical or biological weapons against American or allied forces. This threat was consistent with other deterrent cases in which the United States has asserted that "all options remain on the table." We cannot know the mind of Saddam Hussein in 1991 in order to resolve whether the not-very-disguised threat of possible nuclear first use in response to Iraqi WMD attacks was meaningful to him. The aftermath of war included the discovery that Iraq was closer to developing its own nuclear weapon than experts had previously supposed. Some might use this fact to support the argument that, in 1991 nuclear deterrence *would* have been very meaningful to Saddam Hussein. He obviously believed in the utility of nuclear weapons if he was seeking his own national capability.

On the other hand, there are counterarguments to the necessity for, or the utility of, nuclear deterrence in this case. First, the U.S. and allied coalition had overwhelming superiority in conventional military forces: land, air and sea. In addition, and decisively in the event, the United States had already established itself as the first information age military superpower. Admittedly, Iraq was defeated in a war whose military campaigns were more one sided than even U.S. prewar predictions had been. Nevertheless, Saddam's fate was sealed not only by United States and allied superiority in numbers and kinds of weapons as well as the fighting skills of coalition troops. Iraq was also doomed by the skillful use of U.S. prewar diplomacy that isolated Saddam Hussein from political allies, and therefore from outside military support. Even the Soviet Union, one of Iraq's most important prewar military benefactors, demurred to support the invasion of Kuwait and stood by as the coalition clobbered

Iraq's forces. Meanwhile, the George Bush I administration mobilized political support from leading Arab and Islamic countries, including those opposed to the United States on other issues.

Thus, the first reason why nuclear deterrence might have been less important in Gulf War I than some have contended is the obviously one-sided nature of the diplomatic and military alignments for and against Iraq. The second reason for minimizing instead of maximizing the nuclear role in deterring Iraqi use of WMD is the possibly counterproductive impact of U.S. nuclear use in Iraq. The U.S. plan for Operation Desert Storm was the first generation of "shock and awe" based on decisive air attacks against the brain and central nervous system of the enemy, combined with crushing operational maneuvers against the opponent's field forces. The use of U.S. tactical nuclear weapons in the middle of this campaign would have disrupted the American and coalition air and ground campaigns with the prompt and residual effects of nuclear detonations. Even "precision" use of tactical nukes against suspect sources of chemical and biological weapons could have created no-go zones of imminent danger for U.S. and allied forces, actually delaying closure of ground and air operations and prolonging the war.

Worse would have been the political and psychological impacts of U.S. nuclear first use since Hiroshima: against a non-nuclear weapons state in the Middle East. Despite no love for Saddam Hussein in the region, few Arab and Islamic countries would have been other than appalled that the United States had crossed the nuclear threshold. The anti-Iraq political coalition might have crumbled, insofar as its Arab and Islamic participants were concerned, even before the war had been concluded. In addition, the residual effects of nuclear weapons could spread to Kuwait and Saudi Arabia, the designated beneficiaries of the U.S. and allied military campaign. Nuclear contamination of Saudi Arabia, and especially of its holy cities, would have produced reactions in the Arab and Islamic world that made the later propaganda of al-Qaeda seem tame by comparison.

In addition, win or lose in battle, the introduction of nuclear weapons into Desert Storm might have been unsettling on the American home front. Recall that, even without nuclear use in the Gulf theater of operations, postwar debates over "Gulf war disease" and its possible causes, including troops' exposure to exploding chemical or biological weapons stores, raged for a decade. It is not necessary to enter into the details of those debates or to referee the dispute about probable causes. It is only necessary to appreciate how much worse the situation might be if we were now dealing with the treatment of postconflict nuclear exposure for U.S. and allied troops. After having used nuclear weapons in Desert Storm, the United States might have had a difficult time assembling the next "coalition of the willing" for any mission other than an unprovoked Soviet invasion of NATO.

And speaking of the Soviets – the introduction of U.S. nuclear weapons into Desert Storm could have turned the tables, and not only on Soviet willingness to accept the expulsion of Iraq from Kuwait. The first use of nuclear weapons in anger since 1945, by Americans against a recent Soviet ally, would have created military brushfires under Gorbachev, who was at this very time perched precariously on a tightrope between Cold War hardliners and reformers. Politburo conservatives, using the argument that the United States was now throwing tactical nuclear counterpunches as a matter of routine policy, might have launched the coup of August, 1991 – this time, with success. Anger throughout the Islamic world over the recently concluded Soviet war in Afghanistan would have been submerged by Islamic horror at the nuclear first use by the United States in Iraq. Admittedly these are "counterfactuals," but they are not improbable counterfactuals. The United States, in the aftermath of nuclear use in Iraq, could have ended up with a Soviet Union that remained standing for another decade, combined with an anti-American jihad ten years before 9/11 (perhaps accelerating the date of 9/11 into the first Bush or Clinton presidencies).

The preceding résumé of arguments and examples does not prove that a global system of nuclear primacy cannot ever come to pass. Nor does it prove that, if it did, it would be incompatible with international peace and security. A state in the position of nuclear primacy relative to all others might act as a Leviathan that enforced nuclear abstinence or deterrence upon all others. However, such an outcome, for any state prepared to bet its future on nuclear primacy, would be the triumph of hope over international experience. The complexity of international politics today, and the rapid changes in science and technology that are even now imminent, caution against hegemonial aspirations on the part of any state – with, or without, nuclear weapons. The immediate past century was unkind to empires, and there is no reason to expect nonresistance to imperial ambitions in the twenty-first century. A blatant aspiration for nuclear primacy would draw opposition not only from other nuclear powers, but also from non-nuclear states that feared unilateral domination over their policies. Finally, enforcement of a policy of nuclear primacy would require the nuclear hegemony to repeatedly threaten, or engage in, nuclear war in order to deter conventional aggression – actually raising the costs and risks of war instead of controlling them.

Defense dominance

If the continuation of a regime of deterrence based on assured retaliation is problematical, and if the aspiration for nuclear primacy is self-defeating, a third alternative regime is defense dominance. Defense dominance is popularly associated with the Reagan administration, given President Reagan's call for a research and development program, the Strategic Defense Initiative, to make nuclear weapons and their ballistic delivery vehicles "impotent and obsolete."[5] Reagan's defense initiative had its first public airing in a presidential speech in March, 1983 that took the Pentagon and some of his key political advisors by surprise. Also surprising to some U.S. arms control experts and Reagan advisors were the discussions between Reagan and Soviet leader Mikhail Gorbachev at Reykjavik in 1986, when the two heads of state came within hailing distance of an agreement to eliminate nuclear weapons entirely. Reagan's support for SDI and his seemingly impulsive willingness to put U.S. nuclear weapons on the negotiating table with Russia were less anomalous than they appeared. Reagan was skeptical of a U.S. defense policy based on deterrence and retaliation only, even before arriving at the White House.[6] Briefings after assuming office only added to his incredulity with existing nuclear war plans.

The first term of the Reagan administration was marked by an uptick in U.S.–Soviet tensions. First, the shooting down of Korean Air Lines 007 by Soviet air defenses in the Far East was charged by the United States as a deliberate act of murder and as a political provocation. Second, the Americans (and NATO) were at loggerheads with the Soviets over the NATO decision, taken during the Carter administration, to deploy new generations of ballistic and cruise missiles in Europe beginning in 1983. These NATO "572" deployments were in response to Soviet deployment of new SS-20 intermediate range ballistic missiles targeted on Europe. The Soviets explained their SS-20 deployments as a simple modernization of existing weapons systems, but the United States and NATO regarded the Soviet deployments as an escalation of nuclear danger in Europe.

Third, the general atmosphere of U.S.–Soviet relations became heated by Reagan's call for a U.S. military buildup and his pointed remarks about the Soviet Union being the focus of evil in the world and destined for the "ash heap" of history. Soviet leader Yuri Andropov and others sought to depict the United States as a power bent on aggression and actively planning for a nuclear war – including a possible first strike against the Soviet Union. Soviet intelligence services were ordered to undertake an operation, code named

---

[5] See Donald R. Baucom, *The Origins of SDI, 1944–1983* (Lawrence, Kansas: University Press of Kansas, 1992), for pertinent history of U.S. missile defense programs. On Soviet and Russian missile defenses, see Jennifer G. Mathers, *The Russian Nuclear Shield from Stalin to Yeltsin* (New York: St. Martin's Press, 2000).

[6] Baucom, *The Origins of SDI*, p. 139 and passim. See also: Frances FitzGerald, *Way out there in the Blue: Reagan, Star Wars and the end of the Cold War* (New York: Simon and Schuster, 2000).

RYAN (Russian acronym for nuclear missile surprise attack), of intensive surveillance in the United States and in NATO Europe for timely evidence of plans for prompt nuclear attack. Since Soviet military thinking envisioned that, in the event of a war in Europe between NATO and the Soviet-led Warsaw Pact, the war might grow out of a military exercise by either side, Soviet military and other intelligence collection was especially sensitive to any NATO exercises that rehearsed the delegation of nuclear release authority to operational commanders. As luck would have it, NATO scheduled a nuclear command post exercise, "Able Archer," from November 2 to 11, 1983 that included movement of imaginary NATO forces through all alert phases, from normal readiness to general alert. No real NATO forces were alerted, but alarmist KGB reporting apparently persuaded some officials in Moscow Center that the alerts were real.[7] Although the exercise did not lead to any conflict, President Reagan later learned to his dismay that some Soviet observers were overly impressed by the verisimilitude of Able Archer, given the context of U.S.–Soviet political relations.

Reagan's second term witnessed a thawing in U.S.–Soviet political relations and significant progress in nuclear arms control, including the Intermediate Nuclear Forces (INF) treaty of 1987 that eliminated an entire class of NATO and Soviet medium and intermediate range missiles (ranges from 500 to 5,500 kilometers). Presidents Reagan and Gorbachev acknowledged in a joint communiqué that a nuclear war "could never be won and must never be fought." But they remained apart on the feasibility and the desirability of national missile defenses. Reagan regarded SDI as the escape hatch from an otherwise insoluble dilemma between suicide and surrender in the face of nuclear blackmail. The Soviet political and military leadership saw SDI as part of an American ploy to reduce the credibility of their nuclear deterrent, by denying to Soviet long range missile forces their capacity for assured retaliation after a U.S. first strike. The end of the Cold War and the collapse of the Soviet Union left the matter of U.S. missile defenses, and post-Soviet Russia's reaction to U.S. antimissile defenses, in a state of uncertainty.

The George H.W. Bush administration moved away from the Reagan approach favoring a comprehensive national missile defense system. Instead, President Bush I supported a limited missile defense system against accidental launches and small attacks. This approach that privileged limited compared to comprehensive national defenses was more consistent with the limitations of available technology, which was challenged even to fulfill the less ambitious objective of partial rather than total protection for United States forces or

---

[7] Christopher Andrew and Oleg Gordievsky, *KGB: The Inside Story of Its Foreign Operations from Lenin to Khrushchev* (New York: Harper Perennial, 1991), pp. 599–605.

society. The Clinton administration was rather agnostic on the entire subject of missile defenses. Research, development and testing of BMD technologies continued, but no clear threat perception or sense of urgency pushed missile defenses to the top of the agenda. A Republican majority in Congress pushed through legislation calling for U.S. deployment of a national missile defense system as soon as reliable technology became available. Clinton, like Bush I, left office with the issue of missile defenses open-ended: neither decisively endorsed as a primary military mission, nor rejected as an alternative or complement to offensive nuclear systems. The can was, once again, kicked further down the road.

The George W. Bush administration departed from the low profile policies of Bush I and Clinton with respect to nationwide and global missile defenses. Bush II declared U.S. intent to depart the ABM Treaty and to begin deploying components of a global missile defense system as soon as practicable. Bush national security strategy documents called for a "new triad" of conventional and nuclear offensive strike systems; antimissile defenses; and, third, improved infrastructure to support U.S. defense modernization, including in the area of nuclear weapons. The U.S. Department of Defense guidance identified a need to shift from a "one size fits all" model of deterrence to "tailored deterrence" in which nuclear targeting and force employment would be flexible in order to deal with threats from possible peer competitors, rogue states or even terrorists. Missile defenses would thus fit into a broader framework of offensive and defensive weapons systems, and of adaptive and flexible force planning, which provided for a menu of options that could assure, dissuade, deter or, if necessary, defeat any adversary.[8]

The Bush II national security and nuclear policy guidance seemed to some as if it were a substantial departure from precedent, but the degree of novelty is easily exaggerated. For example, U.S. nuclear targeting policy has been in search of flexibility since the Kennedy administration. Regardless their attitudes toward missile defenses, U.S. Presidents and Defense Secretaries have tasked their military advisors and planners to come up with strike options other than massive counterforce or counter-city attacks. After the Cold War, modern technology enabled the Clinton and Bush II administrations to build cafeteria options and adaptive planning into the nuclear war planning system. The Bush II administration also sought to develop a new generation of low yield and high precision weapons for use against certain targets, including bunkers containing weapons of mass destruction.

---

[8] Amy F. Woolf, *Nuclear Weapons in U.S. National Security Policy: Past, Present, and Prospects* (Washington, D.C.: Congressional Research Service, October 29, 2007), for an informative appraisal with citations to pertinent government documents.

The problem with planning guidance, flexible nuclear command-control systems, and improved or refined options does not lie in the abstract concepts, but in the obduracy of the real world. A "limited" nuclear war is simply a military and humanitarian oxymoron. Defenses cannot change that. Nor can missile defenses overturn the nuclear revolution. The combination of nuclear warheads with ballistic missiles presents any defender with a problem of arithmetic and military complexity. The challenge is not that missile defense technologies will fail to improve, relative to offenses. The challenge with respect to missile defenses is that they will coexist with offenses instead of replacing them. Offenses will continue to appeal to states for deterrence because they are large, fast, and scary. Offenses are inexpensive relative to defenses, and they have acquired the status of acceptability and legitimacy as components of military arsenals that defenses have yet to reach. Perhaps this is a perceptual bias, in favor of offenses and unfair to defenses, but the precedent has been set in the minds of many policy elites and military commentators.

The Bush II program for a global missile defense system included components deployed outside of the United States, including in Eastern Europe.[9] The most controversial of these proposals called for deployment of ten missile interceptors in Poland and a missile warning radar in the Czech Republic.[10] Although approved by NATO foreign ministers in December, 2008, the U.S. plan provoked Russian criticism from the time it was first publicly acknowledged in 2007. Then Russian President Vladimir Putin regarded the deployment of U.S. missile defenses in Eastern Europe as a potential threat to the viability of Russia's nuclear deterrent. Putin and other Russian leaders

---

[9] U.S. Department of Defense, Missile Defense Agency, *Global Ballistic Missile Defense: A Layered Integrated Defense* (Washington, D.C.: BDMS Booklet, Fourth Edition, 2006). For technical assessments, see U.S. Government Accountability Office, *Defense Acquisitions: Status of Ballistic Missile Program in 2004* (Washington, D.C.: GAO, March 2005); and Lisbeth Gronland, et. al., *Technical Realities: An Analysis of the 2004 Deployment of a U.S. Missile Defense System* (Cambridge, Mass.: Union of Concerned Scientists, May 2004). Expert policy discussion and analysis appear in Dean A. Wilkening, *Ballistic-Missile Defence and Strategic Stability* (Oxford: Oxford University Press, 2000).

[10] Dr. Patricia Sanders, Executive Director, Missile Defense Agency, *Missile Defense Program Overview For The European Union, Committee on Foreign Affairs, Subcommittee on Security and Defense* (Washington, D.C.: U.S. Department of Defense, Missile Defense Agency, June 28, 2007), 07-MDA-2623, provides important information about technology and assumptions. Other expert commentary appears in Richard L. Garwin, *Ballistic Missile Defense Deployment in Poland and the Czech Republic*, a talk for the Erice International Seminars, 38th Session, August 21, 2007, RLG2@us.ibm.com or www.fas.org/RLG/. For an expert assessment of Russian perspectives on this issue, see Alexander Pikayev, "Russia and Missile Defences," in Walter Slocombe, Oliver Thranert and Alexander Pikayev, *Does Europe Need a New Missile Defense System?* (Brussels: European Security Forum, November 2007), pp. 19–26.

also saw the proposed deployments as a political provocation intended to showcase U.S. global hegemony. In addition, Russia was already concerned about NATO expansion into former Soviet security space in Russia's "near abroad." Fourth, and finally, Russia rejected the U.S. claim that the missile interceptors and radars were intended to deter or defeat possible missile attacks from Iran or other regional nuclear powers.[11]

Russia's pique was partly driven by the obduracy of its military bureaucracy that continued to resist modernization. Part of the military troglodytes' resistance strategy was to continue NATO and the United States as the prototypical "main enemies" against which Russian military planning had to be benchmarked. From this perspective, Russia continued to need a large mass mobilization army based on conscription, instead of a smaller, more elite force based on voluntary enlistment (and having fewer billets for generals). Putin and his successor as Russian President, Dmitri Medvedev, have indicated awareness of the need for Russia to transition to a military based mainly on contract service, and current Russian plans call for drastic reductions in the size of the armed forces (including the general staff). Regardless the outcome of the political debates over Russian military modernization, retro thinking continues to plague the existing general staff about the intentions of the United States and NATO.

On the other hand, the Bush administration perhaps approached its relationship with Russia, on missile defense and other issues, as if Russia were a petitioner instead of a security partner. Let us suppose a hypothetical-deductive alternative history. Suppose Bush had offered Russia the opportunity to take part in the deployment, management and oversight of a global aerospace defense system. The system could include U.S. missile defense and Russian missile defense and air defense systems – the latter the state of the art. Russian and NATO interceptor, warning and command-control centers would be linked, sharing information in real time and swapping observers from NATO at Russian centers and vice versa. The military to military exchanges

---

[11] For example, see Henry Meyer and Sebastian Alison, "Medvedev Says Russia to Respond to U.S. Missile Deal," *Bloomberg News*, July 9, 2008, in *Johnson's Russia List* 2008 – #129, July 9, 2008, davidjohnson@starpower.net. See also: Associated Press, "Missile defense, Kosovo are 'red lines' for Russia, foreign minister says," *International Herald Tribune Europe*, September 3, 2007, http://www.iht.com/articles/2007/09/03/asia/russia.php. Ellen Barry and Sophia Kishkovsky, "Russia Warns of Missile Deployment," *New York Times*, November 5, 2008, http://www.nytimes.com/2008/11/06/world/europe/06russia.html; and "Russia to deploy missiles near NATO border," Associated Press, November 5, 2008, http://www.msnbc.msn.com/id/27551326/. Kremlin expectations for U.S. cooperation on missile defenses might improve after the departure of the Bush administration in January, 2009. See Simon Tisdall, "Kremlin opts for charm over strong arm on missile defense," *The Guardian*, November 10, 2008, in *Johnson's Russia List* 2008 – #205, November 10, 2008, davidjohnson@starpower.net.

and collaboration would promote trust and help to degrade suspicions based on exaggerations of the system's capabilities or on misperceptions of intent.

Forward thinking Russians and Americans after the Bush II administration could go further. A great opportunity was missed during the 1990s when Russia was left outside of NATO's security community. Russia was dealt with by the U.S. and NATO in the 1990s as a weaker remnant of the former Soviet Union and as a dubious diplomatic partner for building new security architecture. NATO established a postconflict stability operation in Bosnia with Russian participation (albeit not without glitches). But NATO's air war against Serbia in 1999, without United Nations approval, was seen by Russia's leaders as a chest thumping exercise that advertised Russian military weakness. George W. Bush's abrogation of the ABM Treaty and the U.S. decision to begin deploying missile defenses in 2004, although swallowed by Russia without formal objection, seemed also to advertise Russia's lack of any claim to a European security *droit de regard* or even consultative status.

The opportunity still exists for the U.S. and NATO to obtain Russian participation within a larger "northern" security architecture that emphasizes shared threat assessments about terrorism, rogue state attacks, proliferation, and other important security matters. Instead of making missile defenses a wedge issue that raises Russian concerns about NATO enlargement creep and Russian nuclear peril parity, NATO could offer Russia full membership with the understanding that Russia would have to meet NATO's requirements for political democracy and military transparency. Around this new "NATO" (Northern Alliance Treaty Organization), issues such as missile defenses could be restructured or redefined as win-win instead of lose-lose matters. Shared politics would precede shared technology and military to military cooperation. Is this proposal a pipe dream, given the proclivities of the Russian leadership under Putin's de facto or de jure management?

The objective of changing from an offensive-dominant to a defensive-dominant technology environment may be an example of confusing the means with the ends. The experience of the Cold War showed that, with regard to nuclear weapons mated to long range delivery systems, offensive *platforms and technologies* have contributed to defensive *strategies*; that is, nuclear surprise attack does not pay if both sides have a second strike capability. On the other hand, it does not follow that future strategic choices will replicate those of the past – at least, not in detail. In a future in which offensive and defensive systems are mixed into a new brew of dissuasion and deterrence, cognitive simplicity of the Cold War will be replaced by cognitive complexity of the twenty-first century. This offense–defense coexistence, and the complexity that results from it, may demand a rethink of nuclear deterrence even if a nuclear proliferation-constrained world prevails. If, on the other hand, the nonproliferation regime gives way and the spread of nuclear weapons occurs together with missile defense technology breakout, new assessments will be on offer and in demand.

Nuclear abolition and nuclear plenitude

If assured retaliation, nuclear primacy and defense dominance all have shortcomings as candidate paradigms for twenty-first century security, another alternative is nuclear abolition. A sidebar of intermittent interest during the Cold War, abolition has been gathering steam and privileged endorsements from political notables, military experts and others. The appeal of nuclear abolition is obvious and compelling.[12] If nuclear weapons were no longer available to states, then regardless of leaders' intentions and proclivities toward aggression, the option of unacceptable nuclear war would be precluded. The words of scientist Freeman Dyson ring with no less appeal now than they did when he first wrote them during the Reagan administration:

> The abolition of nuclear weapons is a task of the same magnitude as the abolition of slavery. Nuclear weapons are now, as slavery was two hundred years ago, a manifestly evil institution deeply embedded in the structure of our society. People who hope to push the fight against nuclear weapons to a successful conclusion must bring to their task the same qualities which won the fight against slavery: moral conviction, patience, objectivity, and willingness to compromise.[13]

Jonathan Schell has been the most eloquent intellectual voice in favor of nuclear abolition, during and after the Cold War. In his a propos words:

> After the end of the Cold War, the world's nuclear arsenals seemed to have been tamed to a certain extent, but now they are growing and baring their teeth again. Indeed, the bomb is staging a revival, as if to declare: the twenty-first century, like the one before it, belongs to me.[14]

---

[12] A skeptical appraisal of the possibilities for nuclear abolition appears in Lawrence Freedman, "Eliminators, Marginalists and the Politics of Disarmament," Ch. 4 in John Baylis and Robert O'Neill, eds., *Alternative Nuclear Futures: The Role of Nuclear Weapons in the Post-Cold War World* (Oxford: Oxford University Press, 2000), pp. 56–69. For the case that nuclear abolition is both desirable and feasible, see Michael MccGwire, "The Elimination of Nuclear Weapons," Ch. 9 in the same volume, pp. 144–166.

[13] Freeman Dyson, *Weapons and Hope* (New York: Harper Colophon Books, 1985), p. 201.

[14] Jonathan Schell, *The Seventh Decade: The New Shape of Nuclear Danger* (New York: Henry Holt and Company, 2007), pp. 3–4.

Schell's expression of pessimism is prompted by the risk of nuclear weapons spread among states with regional animosities or other inspirations for aggression, and by the alarmist (in his view) tendencies in Bush II national security and nuclear weapons policies. Other writers, in particular Harvard scholar Graham Allison, have emphasized the risk of spreading nuclear weapons and-or technologies falling into the hands of nonstate actors, including terrorists with apocalyptic agendas.[15] Although the Bush II administration certainly exploited public fear of the "mushroom cloud" of nuclear terrorism after 9/11, as part of its preparation for the war in Iraq, the probable connection between more states with nuclear weapons and more opportunities for terrorists to acquire those weapons is clear enough.

One reason for the urgent pessimism on the part of nuclear abolitionists is that the inevitable alternative to nuclear abolition is, in their view, uncontrollable nuclear proliferation and, eventually, nuclear war. Thus it is necessary to discuss in tandem the two regimes of nuclear abolition and nuclear plenitude. Advocates of nuclear abolition see nuclear plenitude, meaning more or less unconstrained proliferation, as unavoidable, unless nuclear weapons are abolished. Controlled proliferation, in which nuclear weapons spread slowly among state actors and predominantly to states that favor the international status quo, is suspect in the minds of abolitionists. But a regime of nuclear plenitude could be one of slow and constrained proliferation as opposed to a manic rat race into Armageddon. States and their leaders, unlike some terrorists, have regimes and societies that they would prefer not to lose.

How optimistic should we be that a regime of nuclear abolition is even possible, given the proclivities of existing nuclear weapons states? On the other hand, how realistic is it to suppose that the nuclear nonproliferation regime can work as well as it has in the past, once the numbers of nuclear weapons states have doubled or tripled? In addition, nuclear weapons are not spreading in a geo-strategically random fashion. The possibility of nuclear proliferation in the Middle East and Asia, among regimes with regional scores to settle and uncertain structures for the control and operation of nuclear forces, is unsettling to expectations about international stability. Already nuclear weapons in the hands of Pakistan and North Korea, not to mention those being sought by Iran, are causing ripples of diplomatic demarches and prompting plans for preemptive military strikes for the purpose of regime change or denuclearization.

One answer to the question "how" with respect to nuclear abolition is to approach the problem incrementally. Nuclear disarmament agreements can

---

[15] Graham Allison, *Nuclear Terrorism: The Ultimate Preventable Catastrophe* (New York: Times Books – Henry Holt and Co., 2004).

move in the direction of smaller forces, leaving a favorable residue of political good will for further reductions and, perhaps eventually, elimination. Even if total abolition of nuclear weapons cannot be obtained, multilateral agreements that provide for qualitative or quantitative arms control or disarmament can be beneficial to peace and security for other reasons. For example, the European Union proposed in December, 2008 to move forward the agendas of arms control and nonproliferation in two steps: (1), a global ban on nuclear testing; and, (2), a moratorium on the production of all fissile material. French President Nicolas Sarkozy wrote to UN Secretary-General Ban Ki-moon December 5 on behalf of the EU that "We are convinced of the necessity to work for general disarmament."[16] The EU also proposed universal ratification of the Comprehensive Test Ban Treaty (CTBT) and further progress in U.S.–Russian negotiations for a follow-on agreement to the 1991 Strategic Arms Reduction Treaty (START), among other measures. The EU proposals might also have been timed to coincide with a conference opening in Paris in early December and sponsored by Global Zero, a group favoring the elimination of nuclear arms worldwide.[17]

Despite increasing international support from political and military leaders and favorable public opinion polls, the case for nuclear abolition, or even for drastic disarmament, meets with intellectual and bureaucratic resistance. American scholar Kenneth N. Waltz, for example, has argued that "more is better" with respect to the spread of nuclear weapons among additional states.[18] Waltz is neither an incurable optimist about the behavior of states, nor a militarist who favors the acquisition of nuclear weapons with the expectation that those forces can be used to prevail in combat. Instead, Waltz has faith in the rationality of state decision makers who face certain destruction from nuclear retaliation after having launched a surprise attack. New nuclear weapons states can be expected to act as carefully with nuclear arsenals as have those who already possess those weapons. The "system" of international interactions, as between prospective participants in nuclear crisis management, pushes both players toward an outcome short of nuclear war.[19]

---

[16] Steven Erlanger, "Europeans Seek to Revive Nuclear Ban," *New York Times*, December 8, 2008, http://www.nytimes.com/2008/12/09/world/europe/09france.html.

[17] Ibid.

[18] For point-counterpoint on this topic, see Scott D. Sagan and Kenneth N. Waltz, *The Spread of Nuclear Weapons: A Debate* (New York: W. W. Norton, 1995).

[19] This author now acknowledges an old debt to colleagues Charles Hermann and Maurice East. During a short but memorable postdoctoral visit to Ohio State University arranged by Allen Millett. Chuck, Mickey and I argued back and forth about international "systems," including what they mean and what they do. The exchanges, which sometimes escalated to rhetorical brawling, were as informative and inspirational of thinking as any books written on the subject. We left the battlefield bloodied but unrepentant.

On the other hand, international systems change, and decision rationalities are very culture bound. The bipolar international nuclear system of the Cold War has given way to something less predictable, and possibly less controlling of rogue behavior, in the present century. For example, in the relationship between India and Pakistan, an attack by Pakistan-based terrorists against Indian Kashmir might escalate into a conventional shooting war, with the latent possibility of escalation to nuclear first use. Note that, in this example, terrorists would not have to obtain nuclear devices in order to contribute to a process of military escalation that ends in a nuclear war. It all depends on politics – always the driver of matters in strategy and in arms control, including disarmament.

The European Union's proposal for incremental disarmament is wise in two ways. First, it recognizes the significance of keeping control of fissile materials – weapons grade uranium and plutonium. The most cost effective way to prevent nuclear terrorism is to prevent terrorists, or states that are willing to support terrorists, from getting access to fissile materials. Existing international stocks of weapons grade uranium and plutonium must be fully accounted for and securely locked down by their hosts. In addition, states seeking to acquire a completed nuclear fuel cycle, allegedly for peaceful purposes, must be open to full throttle inspection by the International Atomic Energy Agency (IAEA). The step from having a complete nuclear fuel cycle to possessing a nuclear weapons capability is a short one in technical terms – but a large one in political and in military significance.

Another opportunity for incrementalism in movement toward a more disarmed world lies in reinvigorating U.S.–Russian nuclear arms reductions. The last two years of the Bush II and Putin presidencies were times of stalemate, punctuated by outbursts of rhetoric from Moscow. Putin still lurks in the wings, but Bush's departure in January, 2009 is permissive of a different U.S. approach that may be compatible with Russian interests. For example, there is neither a present nor a foreseeable need for the United States and Russia to maintain as many deployed strategic nuclear weapons as they now have. Indeed, they could safely go below their agreed limits according to the New START Treaty of 2010.

Having moved forward on New START extension, the U.S. and Russia would then have more credibility in leading others in the nonproliferation parade. Two immediate war stoppers in this category are North Korea and Iran. Russia's participation in the six-party talks designed to roll back North Korea's nuclear capability is important, since Russia and China are North Korea's most strategically important regional neighbors with their own nuclear weapons.[20] With regard to Iran, Russia's past support for Iran's nuclear industry and other economic and military involvements in that country hold potential diplomatic leverage. Intelligence estimates vary about when Iran might actually have a usable nuclear weapon. IAEA nuclear inspectors and some other experts estimated in November, 2009 that Iran had already produced about enough nuclear material, with additional purification, for a single atomic bomb.[21]

Containment of nuclear proliferation in Iran and reversal of North Korean nuclear weapons capabilities are litmus tests for the existing nonproliferation regime. Failure in Tehran and in Pyongyang invites other Middle Eastern and Asian players to the poker table of nuclear stakes. That said, freezing Iran short of deployable nuclear weapons and reversing North Korea's apparent nuclear weapons capability calls for nuanced diplomacy, not preemptive military attack. Bush II national security strategy, with the possibility of conventional or nuclear preemption advertised in declaratory policy mistook the logic of blackmail or intimidation per se for the logic of coercive diplomacy. Coercive diplomacy puts the option of military force at the back end of the cafeteria, emphasizing diplomatic and other nonmilitary incentives for cooperation. The big stick of military power remains on the table as a last resort.

In addition, the policy stories attached to preemptive strikes against Iran and North Korea are not necessarily favorable ones for the United States and its European or Asian allies. Military "success" against North Korea could lead to political upheaval within the DPRK, including overthrow or paralysis of the "Kim family regime" and a splitting of the DPRK military and security forces into competing factions of warlords and criminals. A fragmenting North Korea would threaten stability in South Korea with refugee flows and other chaos. If chaos in North Korea became unmanageable, other states would have to arrange a "stability and security operation" to reestablish

---

[20] Negotiating with North Korea on nuclear weapons or anything else is a roller coaster ride of uncertain destination. For the state of play near the end of the Bush II administration, see "N. Korea nuclear talks fail to break deadlock," Associated Press, December 11, 2008, http//www.msnbc.msn.com/id/28179764.

[21] William J. Broad and David E. Sanger, "Iran Said to Have Fuel for One Weapon," New York Times, November 19, 2008, http://www.nytimes.com/2008/11/20/world/middleeast/20nuke. html. See also: Wisconsin Project on Nuclear Arms Control, Iran's Nuclear Timetable (Washington, D.C.: Iran Watch, November 20, 2008), http://www.iranwatch.org/ourpubs/ articles/irannucleartimetable.html.

order, handle humanitarian crises, and most important for the present discussion, secure control of nuclear weapons, facilities and materials. All this would have to be accomplished by unprecedented cooperation among China (the key player), Russia, South Korea, and the United States – with or without United Nations support. Japan would not, for reasons of history and politics, have a postchaos military footprint in North Korea, but Japan might help to bankroll the stability operation and provide maritime flank security.

In a similar fashion, U.S. or other strikes on Iran intended to disable its nuclear facilities for enrichment or reprocessing might "succeed" in a military fashion – depending on the duration and intensity of the strikes, and the accuracy of United States and allied intelligence. As in North Korea, however, the aftermath of preemptive military attack might be unwelcome political blowback. The "Arab street" and militants throughout the Islamic world, including publics in American allies such as Pakistan and Egypt, would use this attack by the "crusaders" on an Islamic country as a recruiting poster for al-Qaeda and other jihadist groups. With the American military footprint already extended into Iraq (until the end of the year 2011) and Afghanistan (exit undetermined) amid global controversy, further polarization of American relations with selected Middle Eastern and South Asian countries can be expected. The U.S. or other preemptive attackers would also empower the conservatives within Iran, including the ayatollahs with their own military, security and intelligence services, as against the more moderate factions in that country. Iranian support for insurgency and terror against Iraq and Afghanistan would increase, further tying down United States and allied forces in those states and threatening to destabilize their regimes.

If, on the other hand, the United States, Russia and other leading powers in Europe and Asia can work to contain North Korean and Iranian proliferation, then a precedent has been set that supports the nonproliferation regime. This does not lead to nirvana, and perhaps, not even to nuclear abolition. But it does buy time for more progress in arms control and disarmament: denuclearized North Koreans and a non-nuclear Iran reduce the incentives for a nuclear weapons capability in Japan and South Korea, or in Egypt and Saudi Arabia. Progress in disarmament and in nonproliferation, as in all else requiring the voluntary cooperation of governments, comes in discrete steps, not in flashes of epiphany. There is no need to apologize for small victories in politics: just for the failure to try.

What will push nuclear arms control and disarmament the last mile from smaller arsenals to "zero" or effective zero as abolitionists wish is, however, not only the commitment to incremental progress by states. Also required are a compelling moral imperative and, as well, a policy prescription that is strategically sound.[22] States will not be persuaded to disarm themselves for altruistic reasons. They must be convinced that they are more secure in a nuclear-scarce, or nuclear-free, world than in a world in which reliance on nuclear deterrence, nuclear primacy, defense dominance or nuclear anarchy are the guarantors of peace and safety. Hard questions intrude: for example, who will monitor states, in a nuclear-free world, to ensure than no state can execute a breakout into nuclear primacy? The great powers might not trust one another, and international bodies are notoriously lacking in military clout.

Practical obstacles oppose all new undertakings. A new way forward begins with a new idea. For realization of the agenda of nuclear abolitionists, the idea of nuclear weapons must first become as substantively "unthinkable" as genocide has become in the present century. Of course, genocides still occur, within and among states. But they take place outside the legitimacy of international legal and political norms, and in the face of widespread public and media disapproval. Nuclear weapons are a long way from this kind of aura. On account of the "long peace" of the Cold War and the "tradition" of nuclear nonuse in combat since 1945, nuclear weapons are seen by some as deterrent and as pacifiers. And so they have been – under propitious political circumstances. Because politics drives strategy, nuclear weapons in the twentieth century favored stability. In the previous century, nuclear weapons were mostly in the hands of regimes that favored the international status quo, sharing historical and cultural ties that cut across their political disagreements. In the present century, new regimes armed with nukes may have new agendas. For the convenience of the reader, the preceding discussion is followed by Table 1.1 that lists some of the pluses and minuses of each regime.

---

[22] This requires serious study in the discipline of strategy, for which too many political leaders and others in Western societies have a fatal aversion. See Colin S. Gray, *Modern Strategy* (Oxford: Oxford University Press, 1999), esp. Chs. 11–12. According to Sun Tzu: "War is a matter of vital importance to the State; the province of life or death; the road to survival or ruin. It is mandatory that it be thoroughly studied." Sun Tzu, *The Art of War*, translated and with an introduction by Samuel B. Griffith (Oxford: Oxford University Press, 1971), p. 63.

Table 1.1 Pluses and minuses for each of the regimes noted above

	Pluses	Minuses
Deterrence/assured retaliation	Road tested during the Cold War; a lingua franca among those who theorize about or practice strategy and arms control; the idea of assured retaliation is readily understood by amateur politicians and mass publics as well as military specialists	Peace and survival are dependent upon an experiment in applied psychology; states and their leaders are vulnerable to flawed decision making processes, including misperceptions, command-control mishaps, technical glitches and other miscues; deterrence is culture bound
Nuclear primacy	For the primary state, permits extension of its writ by means of coercive threats that are highly credible against weaker nuclear, or non-nuclear powers; a peace of sorts may be imposed by the primary state, via its nuclear Leviathan status	No primary state with exclusive control of nukes can be trusted not to abuse its power; other states will resist domination, and their efforts to increase their power will lead to war; Leviathans all have entropy, and some may give way to anarchy instead of order
Defense dominance	A world of defenses only, or defenses more competent than offenses, could be a world in which nuclear first use or first strike became unacceptable or pointless	Missile defense technologies may lend themselves to "offensive" and destabilizing uses, such as shooting down reconnaissance, warning, communications and navigational satellites

Table 1.1 (Continued) Pluses and minuses for each of the regimes noted above

	Pluses	Minuses
Nuclear abolition	It removes the risk of nuclear war by removing the weapons with which to engage in nuclear combat; nuclear abolition requires political trust that could spill over into reduced numbers of non-nuclear wars; nuclear abolition removes the possibility of destruction of life and value on a hemispheric or global scale	Nuclear abolition may make the world safe, or safer, for the use of non-nuclear forces; nuclear abolition may shift the balance of power toward leaders in advanced technology, non-nuclear weapons; disarmed or destroyed nuclear weapons can be replaced by newer ones – nuclear knowledge cannot be expunged
Nuclear plenitude	Proponents expect that the deterrence "success" of the Cold War would remain in a truly multipolar nuclear system; more states with nuclear weapons will not necessarily be less prudent than their predecessors; nuclear weapons impose their own prudence and restraint on decision makers, regardless of culture	Nuclear plenitude will result in a less manageable and peaceful global system; the system ordering capabilities of states will be overwhelmed by the plurality of decision rationalities and cultures that become nuclear armed; small nuclear wars will become "thinkable" by states and one or more of these wars may escalate into larger, including global, war

2

Cyberwar and nuclear crisis management

Introduction

On largely separate tracks, the U.S. and other governments have continued to maintain and deploy nuclear weapons even as they pursue conventional force modernization based on an information-based Revolution in Military Affairs (RMA).[1] If the ultimate weapons of mass destruction – nuclear weapons – and the supreme weapons of soft power – information warfare – are commingled during a crisis, the product of the two may be an entirely unforeseen and unwelcome hybrid. Crises by definition are exceptional events. No Cold War crisis took place between states armed both with advanced information weapons and with nuclear weapons. But given the durability of the two trends, interest in infowar and in nuclear weapons, the potential for overlap and its implications for nuclear crisis management deserve further study and policy consideration. The discussion below proceeds toward that end, by looking at relevant concepts and examples, some admittedly speculative.

Concepts and definitions

Cyber concepts in plural

The literature and the U.S. government already offer a rich menu of definitions for important cyber related concepts, including cyberspace and cyberpower.

---

[1] The concept of "Revolutions in Military Affairs" is examined with historical case studies in Colin S. Gray, *Strategy for Chaos: Revolutions in Military Affairs and the Evidence of History* (London: Frank Cass, 2002), esp. Ch. 8 on the nuclear RMA. See also: Max Boot, *War Made New: Technology, Warfare, and the Course of History, 1500 to Today* (New York: Gotham Books, 2006), pp. 307–436.

The U.S. armed forces use the umbrella term "computer network operations" consisting of three elements: defense, attack and exploitation (respectively: defending U.S. networks against attack; attacking opposed networks in order to disrupt, deny, degrade or destroy information; and, third, exploiting data taken from the other side's networks and systems.[2] Information warfare can be defined as activities by a state or nonstate actor to exploit the content or processing of information to its advantage in time of peace, crisis, or war, and to deny potential or actual foes the ability to exploit the same means against itself. This is an expansive, and permissive, definition, although it has an inescapable bias toward military- and security-related issues.[3] Information warfare can include both *cyberwar* and *netwar*. Cyberwar, according to John Arquilla and David Ronfeldt, is a comprehensive, information-based approach to battle, normally discussed in terms of high-intensity or mid-intensity

---

[2] Robert A. Miller, Daniel T. Kuehl and Irving Lachow, "Cyber War: Issues in Attack and Defense," *Joint Force Quarterly*, Issue 61, 2nd Quarter 2011, pp. 18–23, esp. pp. 18–19. See also, for pertinent definitions of cyberwar and related concepts: Daniel T. Kuehl, "From Cyberspace to Cyberpower: Defining the Problem," Ch. 2 in Franklin D. Kramer, Stuart H. Starr, and Larry K. Wentz, eds., *Cyberpower and National Security* (Washington, D.C.: National Defense University Press – Potomac Books, Inc., 2009), pp. 24–42. See also, in the same volume: Martin C. Libicki, "Military Cyberpower," Ch. 11, pp. 275–284, and Richard L. Kugler, "Deterrence of Cyber Attacks," Ch. 13, pp. 309–340. Martin C. Libicki, *Cyberdeterrence and Cyberwar* (Santa Monica, Calif.: RAND Corporation, 2009), argues that strategic cyberwar is unlikely to be decisive, although operational cyberwar has an important niche role. Libicki also warns that deterrence in the cyber realm is unlikely to behave as it does in other domains, including conventional war and nuclear deterrence. See also: Will Goodman, "Cyber Deterrence: Tougher in Theory than in Practice?," *Strategic Studies Quarterly*, No. 3 (Fall, 2010), pp. 102–135. Goodman argues that cyberspace poses unique challenges for deterrence but not necessarily impossible ones.

[3] Concepts related to information warfare are discussed in David S. Alberts, John J. Garstka, Richard E. Hayes and David T. Signori, *Understanding Information Age Warfare* (Washington, D.C.: DOD Command and Control Research Program, U.S. Department of Defense, third edition October, 2004), esp. pp. 53–94, and David S. Alberts, John J. Garstka and Frederick P. Stein, *Network Centric Warfare: Developing and Leveraging Information Superiority* (Washington, D.C.: Command and Control Research Program, U.S. Department of Defense, 6th printing April, 2005), esp. pp. 87–122. Col. Thomas X. Hammes, USMC (Ret.), discusses the Pentagon's Joint Publication 3–13, *Information Operations*, and the U.S. Department of Defense understanding of information in modern warfare in Hammes, "Information Warfare," Ch. 4 in G.J. David, Jr. and T.R. McKeldin III, eds., *Ideas as Weapons: Influence and Perception in Modern Warfare* (Washington, D.C.: Potomac Books, 2009), pp. 27–34. See also: John Arquilla, *Worst Enemy: The Reluctant Transformation of the American Military* (Chicago, Ill.: Ivan R. Dee, 2008), esp. Chs. 6–7. For perspective on the role of information operations in Russian military policy, see Timothy L. Thomas, "Russian Information Warfare Theory: The Consequences of August 2008," Ch. 4 in Stephen J. Blank and Richard Weitz, eds., *The Russian Military Today and Tomorrow: Essays in Memory of Mary Fitzgerald* (Carlisle, Pa.: Strategic Studies Institute, U.S. Army War College, July 2010), and Thomas, "Russia's Asymmetrical Approach to Information Warfare," Ch. 5 in Stephen J. Cimbala, ed., *The Russian Military Into the Twenty-first Century* (London: Frank Cass, 2001), pp. 97–121.

conflicts.[4] Netwar is defined by the same authors as a comprehensive, information-based approach to societal conflict. Cyberwar is more the province of states and conventional wars; netwar, more characteristic of nonstate actors and unconventional wars.[5]

U.S. and Russian concepts of information warfare date from the Cold War years, although post-Cold War cyber, communications and electronics technologies have obviously required updating of operational concepts. China's determination to develop an informationized military was partly based on its assessment of U.S. success in Operation Desert Storm in 1991, which demonstrated superiority in technologies for C4ISR (command, control, communications, computers, intelligence, surveillance and reconnaissance), precision strike, and other attributes of advanced technology, information based conventional warfare. Reportedly the People's Liberation Army (PLA) intends to develop a networked C4ISR system as part of its application of "network centric warfare" (NCW) to the Chinese military.[6] Despite some intramural disagreements, PLA expectations include expanded efforts in three operational areas (information operations, military uses of space, and "blue water" naval operations) supported by six types of capability: rapid response; precision strike; information offense and defense; situational awareness; command decision making; and precision support.[7] While experts

---

[4] Richard A. Clarke, former counterterrorism coordinator for the George W. Bush and Clinton administrations, and co-author Robert K. Knake include both cyberwar and netwar activities, as defined by Arquilla and David Ronfeldt, in their concept of "cyber war." See Richard A. Clarke and Robert K. Knake, *Cyber War* (New York: Harper Collins, 2010). For an introduction to this topic, see John Arquilla and David Ronfeldt, "A New Epoch – and Spectrum – of Conflict," in *In Athena's Camp: Preparing for Conflict in the Information Age*, ed. Arquilla and Ronfeldt (Santa Monica, Calif.: RAND, 1997), pp. 1–22. See also, on definitions and concepts of information warfare: Martin Libicki, *What Is Information Warfare?* (Washington: National Defense Univ., ACIS Paper 3, August 1995); Libicki, *Defending Cyberspace and other Metaphors* (Washington: National Defense Univ., Directorate of Advanced Concepts, Technologies, and Information Strategies, February 1997); Arquilla and Ronfeldt, *Cyberwar Is Coming!* (Santa Monica, Calif.: RAND, 1992); David S. Alberts, *The Unintended Consequences of Information Age Technologies: Avoiding the Pitfalls, Seizing the Initiative* (Washington: National Defense Univ., Institute for National Strategic Studies, Center for Advanced Concepts and Technology, April 1996).

[5] Arquilla and Ronfeldt, "The Advent of Netwar," in *In Athena's Camp*, pp. 275–94. With regard to the tasks for U.S. Cyber Command (established in 2009) and its implications for the national security decision making process, see Wesley R. Andrues, "What U.S. Cyber Command Must Do," *Joint Force Quarterly*, Issue 59 (4th Quarter 2010), pp. 115–120.

[6] Some proponents of network centric warfare may have overhyped its possibilities, based on mistaken assumptions about the transferability of business rationality into war. See Milan Vego, "Is the Conduct of War a Business," *Joint Force Quarterly*, Issue 59 (4th Quarter 2010), pp. 57–65.

[7] Kevin Pollpeter, "Towards an Integrative C4ISR System: Informationization and Joint Operations in the People's Liberation Army," Ch. 5 in Roy Kamphausen, David Lai and Andrew Scobell, eds., *The PLA at Home and Abroad: Assessing the Operational Capabilities of China's Military* (Carlisle, Pa.: Strategic Studies Institute, U.S. Army War College, June 2010), pp. 193–235.

note that PLA capabilities for networked operations are still bedeviled by inter-service rivalry as well as needed improvements in technology and doctrine, nevertheless future efforts in these directions will have important implications for U.S. military operations. According to one expert analyst:

> The PLA already possesses or is working on weapons systems that threaten or could threaten the U.S. military. These include advanced air defense systems, especially: "double digit SAMS"; long range cruise missiles; and anti-ship ballistic missiles (ASBM). While each of these systems poses a threat, they become even more formidable when networked to form a system of systems. While much analysis has been done on individual PLA weapons systems, it is C4ISR systems which will facilitate PLA efforts to deny the U.S. military access to a theater.[8]

Presumably, a network centric approach to military operations is a necessary part of any future PRC strategy for "anti-access, area denial" against U.S. or other forces intervening against PRC interests. On the other hand, these operational and tactical advances in information-based warfare do not necessarily summarize the entirety of China's needs or that of any other major power in this domain. Strategic information warfare at the highest level also includes the exploitation of information to deter or defeat attacks against the vital political and military centers of government and the armed forces, and, as well, to protect the critical parts of the civilian infrastructure without which the society and the economy cannot function.

According to Richard A. Clarke and Robert K. Knake, in this regard the United States is probably well ahead of peer competitors and other states in its ability to conduct offensive information warfare. However, the United States is also more dependent upon cyber systems and networks than some peer competitors or other prospective opponents are. Because of this higher network dependency factor and the difficulty of getting privately owned U.S. networks on the same page for cyber defenses, a "cyber war gap" exists that poses a potentially vital threat to U.S. security:

> When you think about "defense" capability and "lack of dependence" together, many nations score far better than the U.S. Their ability to survive a cyber war, with lower costs, compared to what would happen to the U.S., creates a "cyber war gap." They can use cyber war against us and do great damage, while at the same time they may be able to withstand a

---

[8] Ibid., p. 225.

U.S. cyber war response. The existence of that "cyber war gap" may tempt some nation to attack the United States.[9]

In short, the race for strategic as well as operational-tactical military advantage in the information age is well under way. Information-based and information-enhanced sensors, shooter and commanders can turn OODA loops faster than their less info-empowered opponents, convert enemy computer networks into "botnets" of remotely controlled digital zombies, and implant surreptitious software behind enemy (digital) lines, awaiting future activation for the purpose of confusion or destruction. Under the best of circumstances, some of this might even be accomplished entirely in the cyber realm without kinetic overtures or leitmotifs: strategic or operational-tactical surrender, in the face of computer and network paralysis. Although things might turn out this way, the kinetic aspect of warfare, including the use of the most powerful weapons available for mass destruction, cannot be omitted as a consideration to be dealt with, amid the march of cyber tools and the relentless Revolution in Military Affairs (RMA) based on information and electronics.

In this regard, the very concept of "cyberdeterrence" involves degrees of uncertainty and complexity that require a leap of analytic faith beyond what we know, or think we know, about conventional or nuclear deterrence. Cyber attacks generally obscure the identity of the attackers, can be initiated from outside of or within the defender's state territory, are frequently transmitted through third parties without their complicity or knowledge, and can sometimes be repeated almost indefinitely by skilled attackers, even against agile defenders. In addition, the contrast between the principles of cyberdeterrence and nuclear deterrence encourages modesty in the transfer of principles from the latter to the former. As Martin Libicki summarizes:

> In the Cold War nuclear realm, attribution of attack was not a problem; the prospect of battle damage was clear; the 1,000[th] bomb could be as powerful as the first; counterforce was possible; there were no third parties to worry about; private firms were not expected to defend themselves; any hostile nuclear use crossed an acknowledged threshold; no higher levels of war existed; and both sides always had a lot to lose.[10]

---

[9] Clarke and Knake, *Cyber War*, p. 149. The U.S. Defense Advanced Research Projects Agency (DARPA) is reportedly working on new cybersecurity programs potentially capable of learning during an attack and repairing themselves. See Cheryl Pellerin, "DARPA goal for cybersecurity: change the game," American Forces Press Service, December 20, 2010, *http://www.af.mil/news/story_print.asp?id=123235799*

[10] Libicki, *Cyberdeterrence and Cyberwar*, p. xvi.

Although experts might quibble about matters of degree, with respect to some of the preceding points, the case is clearly argued that, compared to nuclear deterrence, cyberdeterrence is a concept in search of further refinement by theoretical elaboration and empirical validation. For example, a division chief for the U.S. Army Global Network Operations Center laments "the lack of any meaningful cyberspace doctrine, or at least a serious consideration of how *cyberspace operations* differs from the closely related *computer network operations*, which is itself a key component of *information operations*." [11]

Airpower theorist and military analyst Benjamin S. Lambeth regards cyberspace as part of the third dimension of warfare that also includes air and space operations. Cyberspace, according to Lambeth, is the "principal domain" in which U.S. air services "exercise their command, control, communications, and ISR (intelligence, surveillance and reconnaissance) capabilities that enable global mobility and rapid long-range strike." [12] In addition, U.S. dominance, or falling behind, in cyberspace has repercussions for U.S. success or failure in aerospace and other domains of conflict:

> Our continued prevalence in cyberspace can help ensure our prevalence in combat operations both within and beyond the atmosphere, which, in turn, will enable our prevalence in overall joint and combined battlespace. On the other side of the coin, any loss of cyberspace dominance on our part can negate our most cherished gains in air and space in virtually an instant. [13]

Added to this is the civil–military interaction that will take place between designated military cyber-samurai and their civilian DOD (and other) superiors in the chain of command who may be cyber-challenged or even pre-cyber in their understanding of information technology and its impacts. The nexus among new information capabilities, their implications for decision making, and their potential vulnerabilities to attack may be comprehended by a select few, if at all.

Crisis management

Crisis management, including nuclear crisis management, is both a competitive and cooperative endeavor between military adversaries. A

---

[11] Andrues, "What U.S. Cyber Command Must Do," p. 115.

[12] Benjamin S. Lambeth, "Airpower, Spacepower, and Cyberpower," *Joint Force Quarterly*, Issue 60, 1st Quarter 2011, pp. 46–53, citation p. 50.

[13] Ibid., p. 51.

crisis is, by definition, a time of great tension and uncertainty.[14] Threats are in the air and time pressure on policy makers seems intense. Each side has objectives that it wants to attain and values that it deems important to protect. During a crisis state behaviors are especially interactive and interdependent with those of another state. It would not be too farfetched to refer to this interdependent stream of interstate crisis behaviors as a system, provided the term "system" is not understood as an entity completely separate from the state or individual behaviors that make it up. The system aspect implies reciprocal causation of the crisis behaviors of "A" by "B," and vice versa.

One aspect of crisis management is the deceptively simple question: what defines a crisis as such? When does the latent capacity of the international order for violence or hostile threat assessment cross over into the terrain of actual crisis behavior? A breakdown of general deterrence in the system raises threat perceptions among various actors, but it does not guarantee that any particular relationship will deteriorate into specific deterrent or compellent threats. Patrick Morgan's concept of "immediate" deterrence failure is useful in defining the onset of a crisis: specific sources of hostile intent have been identified by one state with reference to another, threats have been exchanged, and responses must now be decided upon.[15] The passage into a crisis is equivalent to the shift from Hobbes's world of omnipresent potential for violence to the actual movement of troops and exchanges of diplomatic demarches.

All crises are characterized to some extent by a high degree of threat, short time for decision, and a "fog of crisis" reminiscent of Clausewitz's "fog of war" that confuses crisis participants about what is happening. Before the discipline of crisis management was ever invented by modern scholarship, historians had captured the rush-to-judgment character of much crisis decision

---

[14] For pertinent concepts, see: See Alexander L. George, "A Provisional Theory of Crisis Management," in *Avoiding War: Problems of Crisis Management*, ed. Alexander L. George (Boulder, Colo.: Westview Press, 1991), pp. 22–27, for the political and operational requirements of crisis management; and George, "Strategies for Crisis Management," ibid., pp. 377–94, for descriptions of offensive and defensive crisis management strategies. See also: Ole R. Holsti, "Crisis Decision Making," in *Behavior, Society and Nuclear War*, ed. Philip E. Tetlock, et al. (New York: Oxford Univ. Press, 1989), I, 8–84; and Phil Williams, *Crisis Management* (New York: John Wiley and Sons, 1976). See also Alexander L. George, "Coercive Diplomacy: Definition and Characteristics," in *The Limits of Coercive Diplomacy*, ed. Alexander L. George and William E. Simons (2d ed.; Boulder, Colo.: Westview Press, 1994), esp. pp. 8–9, and in the same volume, Alexander L. George, "The Cuban Missile Crisis: Peaceful Resolution Through Coercive Diplomacy," pp. 111–32.

[15] See Patrick M. Morgan, *Deterrence: A Conceptual Analysis* (Beverly Hills, Calif.: Sage Publications, 1983); and Richard Ned Lebow and Janice Gross Stein, *We All Lost the Cold War* (Princeton, N.J.: Princeton Univ. Press, 1994), pp. 351–55.

making among great powers.[16] The influence of nuclear weapons on crisis decision making is therefore not easy to measure or document because the avoidance of war can be ascribed to many causes. The presence of nuclear forces obviously influences the degree of destruction that can be done should crisis management fail. Short of that catastrophe, the greater interest of scholars is in how the presence of nuclear weapons might affect the decision-making process itself in a crisis. The problem is conceptually elusive: there are so many potentially important causal factors relevant to a decision with regard to war or peace. History is full of dependent variables in search of competing explanations.

Requirements for crisis management

The first requirement of successful crisis management is communications transparency. Transparency includes clear signaling and undistorted communications. Signaling refers to the requirement that each side must send its estimate of the situation to the other. It is not necessary for the two sides to have identical or even initially complementary interests. But a sufficient number of correctly sent and received signals are prerequisite to effective transfer of enemy goals and objectives from one side to the other. If signals are poorly sent or misunderstood, steps taken by the sender or receiver may lead to unintended consequences, including miscalculated escalation.

Communications transparency also includes high fidelity communication between adversaries, and within the respective decision-making structures of each side. High fidelity communication in a crisis can be distorted by everything that might interfere physically, mechanically, or behaviorally with accurate transmission. Electromagnetic pulses that disrupt communication circuitry or physical destruction of communication networks are obvious examples of impediments to high fidelity communication. Cultural differences that prevent accurate understanding of shared meanings between states can confound deterrence as practiced according to one side's theory. As Keith B. Payne notes, with regard to the potential for deterrence failure in the post-Cold War period:

---

[16] For example, see Richard Ned Lebow, *Between Peace and War: The Nature of International Crisis* (Baltimore: Johns Hopkins Univ. Press, 1981); Michael Howard, *Studies in War and Peace* (New York: Viking Press, 1971), pp. 99–109; Gerhard Ritter, *The Schlieffen Plan: Critique of a Myth* (London: Oswald Wolff, 1958); and D.C.B. Lieven, *Russia and the Origins of the First World War* (New York: St. Martin's Press, 1983).

Unfortunately, our expectations of opponents' behavior frequently are unmet, not because our opponents necessarily are irrational but because we do not understand them – their individual values, goals, determination, and commitments – in the context of the engagement, and therefore we are surprised when their "unreasonable" behavior differs from our expectations.[17]

A second requirement of successful crisis management is the reduction of time pressure on policy makers and commanders so that no unintended, provocative steps are taken toward escalation mainly or solely as a result of a misperception that "time is up." Policy makers and military planners are capable of inventing fictive worlds of perception and evaluation in which "H hour" becomes more than a useful benchmark for decision closure. In decision pathologies possible under crisis conditions, deadlines may be confused with policy objectives themselves: ends become means, and means, ends. For example, the war plans of the great powers in July 1914 contributed to a shared self-fulfilling prophecy among leaders in Berlin, St. Petersburg, and Vienna that only by prompt mobilization and attack could decisive losses be avoided in war. Plans predicated on the inflexibility of mobilization timetables proved insufficiently flexible for policy makers who wanted to slow down the momentum of late July and early August toward an irrevocable decision in favor of war.

One result of the compression of decision time in a crisis, compared to typical peacetime patterns, is that the likelihood of Type I (undetected attack) and Type II (falsely detected attack) errors increases. Tactical warning and intelligence networks grow accustomed to the routine behavior of other state forces and may misinterpret non-routine behavior. Unexpected surges in alert levels or uncharacteristic deployment patterns could trigger misreadings of indicators by tactical operators. As Bruce G. Blair has argued:

> In fact, one distinguishing feature of a crisis is its murkiness. By definition, the Type I and Type II error rates of the intelligence and warning systems rapidly degrade. A crisis not only ushers in the proverbial fog of crisis symptomatic of error-prone strategic warning but also ushers in a fog of battle arising from an analogous deterioration of tactical warning.[18]

---

[17] Keith B. Payne, *Deterrence in the Second Nuclear Age* (Lexington, Ky: Univ. Press of Kentucky, 1996), p. 57. See also David Jablonsky, *Strategic Rationality Is Not Enough: Hitler and the Concept of Crazy States* (Carlisle Barracks, Pa.: USAWC, Strategic Studies Institute, 8 August 1991), esp. pp. 5–8 and pp. 31–37.

[18] Bruce G. Blair, *The Logic of Accidental Nuclear War* (Washington: Brookings Institution, 1993), p. 237.

A third attribute of successful crisis management is that each side should be able to offer the other a safety valve or a face-saving exit from a predicament that has escalated beyond its original expectations. The search for options should back neither crisis participant into a corner from which there is no graceful retreat. For example, during the Cuban missile crisis of 1962, President Kennedy was able to offer Soviet Premier Khrushchev a face-saving exit from his overextended missile deployments. Kennedy publicly committed the United States to refrain from future military aggression against Cuba and privately agreed to remove and dismantle Jupiter medium-range ballistic missiles previously deployed among US NATO allies.[19] Kennedy and his inner circle recognized, after some days of deliberation and clearer focus on the Soviet view of events, that the United States would lose, not gain, by a public humiliation of Khrushchev that might, in turn, diminish Khrushchev's interest in any mutually agreed solution to the crisis.

A fourth attribute of successful crisis management is that each side maintains an accurate perception of the other side's intentions and military capabilities. This becomes difficult during a crisis because, in the heat of a partly competitive relationship and a threat-intensive environment, intentions and capabilities can change. Robert Jervis warned that Cold War beliefs in the inevitability of war might have created a self-fulfilling prophecy:

> The superpowers' beliefs about whether or not war between them is inevitable create reality as much as they reflect it. Because preemption could be the only rational reason to launch an all-out war, beliefs about what the other side is about to do are of major importance and depend in large part on an estimate of the other's beliefs about what the first side will do.[20]

Intentions can change during a crisis if policy makers become more optimistic about gains or more pessimistic about potential losses during the crisis. Capabilities can change due to the management of military alerts and the deployment or other movement of military forces. Heightened states of military readiness on each side are intended to send a two-sided signal: of readiness for the worst if the other side attacks, and of a non-threatening steadiness of purpose in the face of enemy passivity. This mixed message is hard to send under the best of crisis management conditions, since each

---

[19] Lebow and Stein, *We All Lost the Cold War*, pp. 122–23.

[20] Robert Jervis, *The Meaning of the Nuclear Revolution: Statecraft and the Prospect of Armageddon* (Ithaca, N.Y.: Cornell Univ. Press, 1989), p. 183.

state's behaviors and communications, as observed by its opponent, may not seem consistent. Under the stress of time pressures and of military threats, different parts of complex security organizations may be making decisions from the perspective of their narrowly defined, bureaucratic interests. These bureaucratically chosen decisions and actions may not coincide with the policy makers' intent, nor with the decisions and actions of other parts of the government. As Alexander L. George has explained:

> It is important to recognize that the ability of top-level political authorities to maintain control over the moves and actions of military forces is made difficult because of the exceedingly large number of often complex standing orders that come into effect at the onset of a crisis and as it intensifies. It is not easy for top-level political authorities to have full and timely knowledge of the multitude of existing standing orders. As a result, they may fail to coordinate some critically important standing orders with their overall crisis management strategy.[21]

As policy makers may be challenged to control numerous and diverse standard operating procedures, political leaders may also be insufficiently sensitive to the costs of sudden changes in standing orders or unaware of the rationale underlying those orders. For example, heads of state or government may not be aware that more permissive rules of engagement for military forces operating in harm's way come into play once higher levels of alert have been authorized.[22]

Information warfare and nuclear crisis management

Information warfare has the potential to attack or to disrupt successful crisis management on each of the preceding attributes. First, information warfare can muddy the signals being sent from one side to the other in a crisis. This can be done deliberately or inadvertently. Suppose one side plants a virus or worm in the other's communications networks.[23] The virus or worm becomes

---

[21] Alexander L. George, "The Tension Between 'Military Logic' and Requirements of Diplomacy in Crisis Management," in *Avoiding War: Problems of Crisis Management*, pp. 13–21, citation p. 18.

[22] Ibid.

[23] A virus is a self-replicating program intended to destroy or alter the contents of other files stored on floppy disks or hard drives. Worms corrupt the integrity of software and information systems from the "inside out" in ways that create weaknesses exploitable by an enemy.

activated during the crisis and destroys or alters information. The missing or altered information may make it more difficult for the cyber-victim to arrange a military attack. But destroyed or altered information may mislead either side into thinking that its signal has been correctly interpreted when it has not. Thus, side A may intend to signal "resolve" instead of "yield" to its opponent on a particular issue. Side B, misperceiving a "yield" message, may decide to continue its aggression, meeting unexpected resistance and causing a much more dangerous situation to develop.

Infowar can also destroy or disrupt communication channels necessary for successful crisis management. One way infowar can do this is to disrupt communication links between policy makers and military commanders during a period of high threat and severe time pressure. Two kinds of unanticipated problems, from the standpoint of civil-military relations, are possible under these conditions. First, political leaders may have predelegated limited authority for nuclear release or launch under restrictive conditions: only when these few conditions obtain, according to the protocols of predelegation, would military commanders be authorized to employ nuclear weapons distributed within their command. Clogged, destroyed, or disrupted communications could prevent top leaders from knowing that military commanders perceived a situation to be far more desperate, and thus permissive of nuclear initiative, than it really was. For example, during the Cold War, disrupted communications between the U.S. National Command Authority and ballistic missile submarines, once the latter came under attack, could have resulted in a joint decision by submarine officers and crew to launch in the absence of contrary instructions.

Second, information warfare during a crisis will almost certainly increase the time pressure under which political leaders operate. It may do this literally, or it may affect the perceived time lines within which the policy making process can make its decisions. Once either side sees parts of its command, control, and communications system being subverted by phony information or extraneous cyber-noise, its sense of panic at the possible loss of military options will be enormous. In the case of U.S. Cold War nuclear war plans, for example, disruption of even portions of the strategic command, control, and communications system could have prevented competent execution of parts of the SIOP (the strategic nuclear war plan). The SIOP depended upon finely orchestrated time-on-target estimates and precise damage expectancies against various classes of targets. Partially misinformed or disinformed networks and communications centers would have led to redundant attacks against the same target sets and, quite possibly, unplanned attacks on friendly military or civilian installations.

A third potentially disruptive effect of infowar on nuclear crisis management is that infowar may reduce the search for available alternatives to the few and desperate. Policy makers searching for escapes from crisis denouements need flexible options and creative problem-solving. Victims of information warfare may have a diminished ability to solve problems routinely, let alone creatively, once information networks are filled with flotsam and jetsam. Questions to operators will be poorly posed, and responses (if available at all) will be driven toward the least common denominator of previously programmed standard operating procedures. Retaliatory systems that depend on launch-on-warning instead of survival after riding out an attack are especially vulnerable to reduced time cycles and restricted alternatives:

> A well-designed warning system cannot save commanders from misjudging the situation under the constraints of time and information imposed by a posture of launch on warning. Such a posture truncates the decision process too early for iterative estimates to converge on reality. Rapid reaction is inherently unstable because it cuts short the learning time needed to match perception with reality.[24]

The propensity to search for the first available alternative that meets minimum satisfactory conditions of goal attainment is strong enough under normal conditions in nonmilitary bureaucratic organizations.[25] In civil–military command and control systems under the stress of nuclear crisis decision making, the first available alternative may quite literally be the last. Or, so policy makers and their military advisors may persuade themselves. Accordingly, the bias toward prompt and adequate solutions is strong. During the Cuban missile crisis, for example, a number of members of the presidential advisory group continued to propound an air strike and invasion of Cuba during the entire 13 days of crisis deliberation. Had less time been available for debate and had President Kennedy not deliberately structured the discussion in a way that forced alternatives to the surface, the air strike and invasion might well have been the chosen alternative.[26]

Fourth and finally on the issue of crisis management, infowar can cause flawed images of each side's intentions and capabilities to be conveyed to the other, with potentially disastrous results. Another example from the Cuban missile crisis demonstrates the possible side effects of simple misunderstanding and noncommunication on U.S. crisis management. At the

---

[24] Blair, *The Logic of Accidental Nuclear War*, p. 252.

[25] James G. March and Herbert A. Simon, *Organizations* (New York: John Wiley and Sons, 1958), pp. 140, 146.

[26] Lebow and Stein, *We All Lost the Cold War*, pp. 335–36.

most tense period of the crisis, a U-2 reconnaissance aircraft got off course and strayed into Soviet airspace. U.S. and Soviet fighters scrambled, and a possible Arctic confrontation of air forces loomed. Khrushchev later told Kennedy that Soviet air defenses might have interpreted the U-2 flight as a prestrike reconnaissance mission or as a bomber, calling for a compensatory response by Moscow.[27] Fortunately Moscow chose to give the United States the benefit of the doubt in this instance and to permit U.S. fighters to escort the wayward U-2 back to Alaska. Why this scheduled U-2 mission was not scrubbed once the crisis began has never been fully revealed; the answer may be as simple as bureaucratic inertia compounded by noncommunication down the chain of command by policy makers who failed to appreciate the risk of "normal" reconnaissance under these extraordinary conditions.

Other implications

The outcome of a nuclear crisis management scenario influenced by information operations may not be a favorable one. Despite the best efforts of crisis participants, the dispute may degenerate into a nuclear first use or first strike by one side and retaliation by the other. In that situation, information operations by either, or both, sides might make it more difficult to limit the war and bring it to a conclusion before catastrophic destruction and loss of life had taken place. Although there are no such things as "small" nuclear wars, compared to conventional wars, there can be different kinds of "nuclear" wars, in terms of their proximate causes and consequences.[28] Possibilities include: a nuclear attack from an unknown source; an ambiguous case of possible, but not proved, nuclear first use; a nuclear "test" detonation intended to intimidate but with no immediate destruction; or, a conventional strike mistaken at least initially for a nuclear one. As George H. Quester has noted:

> The United States and other powers have developed some very large and powerful conventional warheads, intended for destroying the hardened underground bunkers that may house an enemy command post or a

---

[27] Graham T. Allison, *Essence of Decision: Explaining the Cuban Missile Crisis* (Boston: Little, Brown, 1971), p. 141. See also Scott D. Sagan, *Moving Targets: Nuclear Strategy and National Security* (Princeton, N.J.: Princeton Univ. Press, 1989), p. 147; and Lebow and Stein, *We All Lost the Cold War*, p. 342.

[28] For pertinent scenarios, see George H. Quester, *Nuclear First Strike: Consequences of a Broken Taboo* (Baltimore, Md.: Johns Hopkins University Press, 2006), pp. 24–52.

hard-sheltered weapons system. Such "bunker-buster" bombs radiate a sound signal when they are used and an underground seismic signal that could be mistaken from a distance for the signature of a small nuclear warhead.[29]

The dominant scenario of a general nuclear war between the United States and the Soviet Union preoccupied Cold War policymakers and, under that assumption, concerns about escalation control and war termination were swamped by apocalyptic visions of the end of days. The second nuclear age, roughly coinciding with the end of the Cold War and the demise of the Soviet Union, offers a more complicated menu of nuclear possibilities and responses.[30] Interest in the threat or use of nuclear weapons by rogue states, by aspiring regional hegemons or by terrorists, abetted by the possible spread of nuclear weapons among currently non-nuclear weapons states, stretches the ingenuity of military planners and fiction writers.

In addition to the world's worst characters engaged in nuclear threat or first use, there is also the possibility of backsliding in political conditions as between the United States and Russia, or Russia and China, or China and India (among current nuclear weapons states). The nuclear "establishment" or P-5 thus includes cases of current debellicism or pacification that depend upon the continuation of favorable political auguries in regional or global politics. Politically unthinkable conflicts of one decade have a way of evolving into the politically unavoidable wars of another – World War I is instructive in this regard. The war between Russia and Georgia in August, 2008 was a reminder that local conflicts on regional fault lines between blocs or major powers have the potential to expand into worse. So, too, were the Balkan wars of Yugoslav succession in the 1990s. In these cases, Russia's one-sided military advantage relative to Georgia in 2008, and NATO's military power relative to that of Bosnians of all stripes in 1995 and Serbia in 1999, contributed to war termination without further international escalation.

---

[29] Ibid., p. 27.

[30] Assessments of deterrence before and after the Cold War appear in: Patrick M. Morgan, *Deterrence Now* (Cambridge: Cambridge University Press, 2003); Colin S. Gray, *The Second Nuclear Age* (Boulder, Colo.: Lynne Rienner, 1999); Keith B. Payne, *Deterrence in the Second Nuclear Age* (Lexington, Ky.: University Press of Kentucky, 1996); Robert Jervis, *The Meaning of the Nuclear Revolution: Statecraft and the Prospect of Armageddon* (Ithaca, N.Y.: Cornell University Press, 1989); and Lawrence Freedman, *The Evolution of Nuclear Strategy* (New York: St. Martin's Press, 1981 and 1983. Michael Krepon emphasizes that deterrence in the first nuclear age "worked," to the extent that it did so, only in conjunction with containment, diplomacy, military strength and arms control. See Krepon, *Better Safe than Sorry: The Ironies of Living with the Bomb* (Stanford, Calif.: Stanford University Press, 2009), passim.

Escalation of a conventional war into nuclear first use remains possible where operational or tactical nuclear weapons have been deployed with national or coalition armed forces. In allied NATO territory, the U.S. deploys several hundred substrategic, air delivered nuclear weapons among bases in Belgium, Germany, Italy, the Netherlands, and Turkey.[31] Russia probably retains several thousands of operational or tactical nuclear weapons, including significant numbers deployed in western Russia.[32] The New START agreement, once ratified, establishes a notional parity between the United States and Russia in nuclear systems of intercontinental range.[33] But U.S. and allied NATO superiority in advanced technology, information-based conventional military power leaves Russia heavily reliant on tactical nukes as compensation for comparative weakness in non-nuclear forces. NATO's capitals breathed a sigh of relief when Russia's officially approved Military Doctrine of 2010 did not seem to lower the bar for nuclear first use, compared to previous editions.[34]

Russia's military doctrine indicates a willingness to engage in nuclear first use in situations of extreme urgency for Russia, as defined by its political leadership.[35] And, despite evident superiority in conventional forces relative to those of Russia, neither the United States nor NATO is necessarily eager to get rid of their remaining substrategic nukes deployed among American NATO allies. An expert panel convened by NATO to set the stage for its 2010 review of the alliance's military doctrine was carefully ambivalent on the issue of the alliance's forward deployed nuclear weapons. The issue of negotiating away these weapons in return for parallel concessions from Russia was left open for

---

[31] For detailed information on U.S. tactical nuclear weapons deployed in Europe, see Hans M. Kristensen, *U.S. Nuclear Weapons in Europe: A Review of Post-Cold War Policy, Force Levels, and War Planning* (Washington, D.C.: Natural Resources Defense Council, February 2005).

[32] See Pavel Podvig, "What to do about tactical nuclear weapons," *Bulletin of the Atomic Scientists*, February 25, 2010, http://thebulletin.org, in *Johnson's Russia List* 2010 – #43, March 3, 2010, davidjohnson@starpower.net, and Jacob W. Kipp, "Russia's Tactical Nuclear Weapons and Eurasian Security," *Jamestown Foundation Eurasia Defense Monitor*, March 5, 2010, in *Johnson's Russia List* 2010 – #46, March 8, 2010, davidjohnson@starpower.net, for pertinent insights and analysis.

[33] *Treaty between the United States of America and the Russian Federation on Measures for the Further Reduction and Limitation of Strategic Offensive Arms* (Washington, D.C.: U.S. Department of State, April 8, 2010), http://www.state.gov/documents/organization/140035.pdf.

[34] Text, "The Military Doctrine of the Russian Federation," www.Kremlin.ru, February 5, 2010, in *Johnson's Russia List* 2010 – #35, February 19, 2010, davidjohnson@starpower.net. See also: Nikolai Sokov, "The New, 2010 Russian Military Doctrine: The Nuclear Angle," Center for Nonproliferation Studies, Monterey Institute of International Studies, February 5, 2010, http://cns.miis.edu/stories/100205_russian_nuclear_doctrine.htm.

[35] See the analysis by Keir Giles, *The Military Doctrine of the Russian Federation 2010*, NATO *Research Review* (Rome: NATO Defense College, Research Division, February 2010), esp. pp. 1–2 and 5–6.

further discussion. On the other hand, the NATO expert report underscored the present majority sentiment of governments that these weapons provided a necessary link in the chain of alliance deterrence options.[36]

Imagine now the unfolding of a nuclear crisis or the taking of a decision for nuclear first use, under the conditions of both NATO and Russian campaigns employing strategic disinformation and information operations intended to disrupt opposed command-control, communications and warning systems. Disruptive information operations against enemy systems on the threshold of nuclear first use, or shortly thereafter, could increase the already substantial difficulty of bringing fighting to a halt before a European-wide theater conflict or a strategic nuclear war. All of the previously cited difficulties in crisis management under the shadow of nuclear deterrence pending a decision for first use would be compounded by additional uncertainty and friction after the nuclear threshold had been crossed.

Three new kinds of frictions would be posed for NATO. The cohesion of allied governments would be tested under conditions of unprecedented stress and danger, doubtless aided by a confused situation on the field of battle. Second, reliable intelligence about Russian intentions following Russian or NATO first use would be essential, but challenging to nail down. Third, the first use of a nuclear weapon in anger since Nagasaki would establish a new psychological, political and moral universe within which negotiators for de-escalation and war termination would somehow have to maintain their sangfroid, obtain agreed stand-downs from their militaries, and return nuclear capable launchers and weapons to secured, but transparent, locations. All of this would be taking place within the panic-spreading capabilities of 24/7 news networks and the internet.

A "man from Mars" might well say: OK, if information operations get in the way of de-escalation, then omit them. But the political desire to do so is in conflict with the military necessity for timely information gathering, assessment, and penetration of enemy networks – in order to accomplish two necessary, but somewhat opposed, missions. First, each side would want to anticipate correctly the timing and character of the other's decision for nuclear first use – and, if possible, to throw logic bombs, Trojan horses, electronic warfare, or other impediments in the way (or, if finesse is not at hand, bombing the relevant installations is always an option, although an obviously provocative one). The second, and somewhat opposed, mission is to communicate reliably with the other side one's preference for de-

---

[36] *NATO 2020: Assured Security; Dynamic Engagement, Analysis and Recommendations of the Group of Experts on a New Strategic Concept for NATO* (Brussels: North Atlantic Treaty Organization, May 17, 2010), pp. 43–44.

escalation, one's willingness to do so if reciprocity can be obtained, and one's awareness of the possibility that the situation will shortly get out of hand. Consider the Russian General Staff and President's office filtering this hydra-headed group of messages while forces were grappling in Georgia or Ukraine (having been taken into NATO membership the previous year, over Russia's objections).

The problem of nuanced messages and the management of de-escalation, even short of war, was illustrated by NATO's command post exercise Able Archer in November, 1983. Able Archer was intended to simulate, among other things, the taking of an alliance decision for nuclear first use. However, some in the Russian intelligence services apparently mistook this exercise for the real thing – or, they judged, it was close enough to merit pushing their alarmism up the chain of command to Moscow Central. Russian sensitivities to the possibility of U.S. or NATO nuclear first use were high at this time, partly due to NATO's decision to begin deploying Pershing II ballistic missiles and ground-launched cruise missiles of intermediate range in Europe, beginning in the fall of 1983. Although forces were not alerted and no immediate nuclear scare captured the imagination of leaders in Washington, Moscow or Brussels, the case illustrates how mistaken interpretations of "normal" events can overvalue pessimistic assessment at just the wrong time.[37]

Similar problems in coordinating the management of de-escalation and conflict termination with the conduct of information operations might appear in two other situations. First, already alluded to, is the use of a bunker-busting or other advanced technology conventional weapon that the other side, during the fog of crisis or war, confused with a nuclear first use or first strike. Russia expressed this concern specifically during New START negotiations in 2010, with regard to American plans to deploy some conventionally armed ballistic missiles on nuclear capable intercontinental or transoceanic launchers. New START counting rules will regard conventionally armed ballistic missiles as also nuclear capable launchers and, therefore, subject to overall restrictions on the numbers of deployed launchers and weapons. U.S. plans for Prompt Global Strike (PGS) systems including missiles or future space planes were

---

[37] On Able Archer and its implications, see Christopher Andrew and Oleg Gordievsky, *KGB: The Inside Story of Its Foreign Operations from Lenin to Gorbachev* (New York: Harper Perennial, 1991), pp. 599–601. Additional parts of the background relevant to political tensions at this time included U.S.-announced plans for the Strategic Defensive Initiative (SDI) in the spring of 1983, the KAL 007 shootdown by a Soviet fighter in September, 1983 and an ongoing KGB–GRU intelligence operation (RYAN) to detect telltale signs of any U.S. or NATO decision for a nuclear attack (Ibid., pp. 582–598). See also, on Able Archer and other components of a "war scare" atmosphere during this time: David E. Hoffman, *The Dead Hand: The Untold Story of the Cold War Arms Race and Its Dangerous Legacy* (New York: Doubleday, 2009), pp. 72–98.

first approved during the George W. Bush administration and carried forward under the Obama administration.[38]

A second illustration, apart from escalation in Europe, of the problem of managing escalation control and conflict termination along with information operations, is provided by the possibility of a joint NATO-Russian theater missile defense (possibly including air defenses) system. The idea has expert and highly visible political proponents on both sides of the Atlantic, and official Russian commentators have not closed the door to the possibility of some cooperation on ballistic missile defenses (BMD). Here NATO and Russia are facing in two political directions: wariness, but also openness, toward one another; and, second, concern about possible future Iranian or other Middle Eastern nuclear weapons in the hands of leaders beyond deterrence based on the credible threat of nuclear (or other) retaliation.

However, the problems of obtaining missile defense cooperation as between NATO and Russia are not only political. Even with the best of intentions among United States, NATO and Russian negotiators, the military–technical problems of coordinating BMD command-control and communications systems are considerable. Indeed, they are not strictly "military-technical" but also heavily embedded with issues of political sovereignty, classified intelligence, and trust, among governments and militaries. Even the militaries among NATO members differ as to their national traditions, military service identities, experiences in nuclear arms control, and willingness to share online information in real times with temporary partners who may be future enemies. For example, if a European theater-wide system of intelligence and missile attack warning is established, how many capitals will host relevant servers and receive timely output? Who will decide that a missile warning is now a threat requiring activation of the European BMD system – can a single nation do so if a missile is headed its way, or must NATO (including the United States) and Russia agree before taking responsive action?

If a political crisis as between NATO and Russia erupts after a cooperative BMD system has been established, will Russian or American cyberwarriors attempt to spoof or otherwise negate the other's missile defense component? Would it be better to reassure Russia as to the surety of its individually

---

[38] Thus Russian nervousness about U.S. missile defense deployments in Europe is not only related to the possible deployment of those systems without consultation or participation by Russia. Russia also takes into account U.S. evolving capabilities for conventional deep strike and cyber warfare, which in combination with American long range nuclear forces and missile defenses, might (according to Russian pessimists) diminish or nullify Russia's options for response. See Stephen J. Blank, *Russia and Arms Control: Are There Opportunities for the Obama Administration?* (Carlisle, Pa.: Strategic Studies Institute, U.S. Army War College, March 2009), pp. 107–118.

based, or shared-with-NATO, missile defenses, as against the possibility of a conventional or nuclear preemption? Neither Russia nor the United States will want to relinquish sovereign control over its part of any cooperative missile defenses. However, would it be prudent to announce a withdrawal from the cooperative aspect of the regional BMD system during a crisis, or to maintain the fiction of cooperation while attacking the other side's cyber systems with Trojan horses, logic bombs and trap doors – just in case? Perhaps, in future nuclear or other crises, the U.S. and Russian cyber commands should have their own direct "hot line" – or, in this case, encrypted digital link.

Conclusion

The possible combination of information warfare with continuing nuclear deterrence after the Cold War could have unintended by-products, and these may be dangerous for stability. Optimistic expectations about the use of information warfare to defeat or disrupt opponents on the conventional, high-technology battlefield – in cases where nuclear complications do not figure – may be justified. On the other hand, where the shadow of possible nuclear deterrence failure hangs over the decision-making process between or among states in conflict, the infowarriors' efforts to obtain dominant battlespace knowledge may provoke the opponent instead of deterring it. As scholars and policy analysts Keir A. Lieber and Daryl G. Press, have noted, with respect to U.S. superior performance at the sharp end of the conventional RMA: "A central strategic puzzle of modern war is that the tactics best suited to dominating the conventional battlefield are the same ones most likely to trigger nuclear escalation."[39]

The objective of infowar in conventional warfare is to deny enemy forces battlespace awareness and to obtain dominant awareness for oneself, as the United States largely was able to do in the Gulf War of 1991.[40] In a crisis with nuclear weapons available to the side against which infowar is used, crippling the foe's intelligence and command and control systems is an objective possibly at variance with controlling conflict and prevailing at an acceptable cost. And under some conditions of nuclear crisis management, crippling the

---

[39] Keir A. Lieber and Daryl G. Press, "The Nukes We Need; Preserving the American Deterrent," *Foreign Affairs*, No. 6 (November/December 2009), pp. 39–51, citation p. 43.

[40] As David Alberts points out, "Information dominance would be of only academic interest, if we could not turn this information dominance into battlefield dominance." See Alberts, "The Future of Command and Control with DBK," in *Dominant Battlespace Knowledge*, ed. Stuart E. Johnson and Martin C. Libicki (Washington: National Defense Univ., 1996), pp. 77–102, citation p. 80.

C4ISR of the foe may be self-defeating. Deterrence, whether it is based on the credible threat of denial or retaliation, must be successfully communicated to – and believed by – the other side.[41] Whether nuclear or other deterrence can work in a particular context is more dependent upon political, as opposed to military, variables. As Lawrence Freedman has noted:

> What is often forgotten in strategic studies, preoccupied with military capabilities, is that the balance of terror rests upon a particular arrangement of political relations as much as on the quantity and quality of the respective nuclear arsenals. Movements in these political relations can prove far more disturbing to nuclear stability than any movements of purely military factors.[42]

---

[41] As Colin S. Gray has noted, "Because deterrence flows from a relationship, it cannot reside in unilateral capabilities, behavior or intentions. Anyone who refers to *the* deterrent policy plainly does not understand the subject." Gray, *Explorations in Strategy* (Westport, Conn.: Greenwood Press, 1996), p. 33.

[42] Lawrence Freedman, *The Evolution of Nuclear Strategy*, Third Edition (New York: Palgrave Macmillan, 2003), p. 463.

3

Geography and nuclear arms control

Introduction

Geography and nuclear arms control might seem an odd match in subject matter. But there are a number of aspects of nuclear arms control, including proliferation, arms reductions and defenses, that are embedded in geostrategic assumptions and perspectives. The following discussion identifies some of the geostrategic contexts within which nuclear deterrence, war planning and arms control must take place. As Professor Colin Gray has noted: "The influence of geography upon the character of conflict is pervasive at all levels of analysis: policy, grand strategy, military strategy, tactics, and technological choices and performance."[1]

Geography, including geopolitics and geostrategy, is among the fundamental contexts within which strategy making takes place: political, social–cultural, economic, technological, military-strategic, geopolitical and geostrategic, and historical.[2] One might, in deference to Sun Tzu, add moral–psychological to the list of important contexts for military affairs, although aspects of both are

---

[1] Colin S. Gray, "Geography and grand strategy," Ch. 9 in Gray, *Strategy and History: Essays on theory and practice* (London: Routledge, 2006), pp. 137–150, citation p. 147. Gray's discussion of the relationship between geography and strategy includes political as well as physical geography and the impact of geography on strategic culture, with pertinent references to "classics" and modern studies. The significance of geography, geopolitics and geostrategic thinking is also explicit in numerous contemporary works about strategy and global politics – including those with other messages. See, for example, Thomas P.M. Barnett, *The Pentagon's New Map: War and Peace in the Twenty-first Century* (New York: Berkley Books, 2004).

[2] Colin S. Gray, *Fighting Talk: Forty Maxims on War, Peace, and Strategy* (Westport, Ct.: Praeger Security International, 2007), pp. 3–6 and 78–81.

included in the political, historical and socio–cultural milieus.[3] It follows that the relationship between geopolitics and geostrategy, on the one hand, and nuclear weapons or nuclear arms control, on the other, is also embedded in these various contexts for strategy.

The nuclear revolution cannot escape geography – and vice versa

At the most basic level, geography and its evident facts, including distance, topography, weather and climate, and terrain, cannot be escaped. Some geopolitical theorists of the nineteenth and twentieth centuries may have overstated a good case for the relevancy of geostrategy. Nevertheless, real military planners must plan against the obdurate facts of the moment, and these facts include geography. For example, geography, including climate, is among the reasons for Russia's history of resisting conquest successfully. Any state territory that can resist the predations of Charles XII, Napoleon and Hitler must present formidable natural barriers, apart from the fighting power of its armed forces or the skill of its commanders. On the other hand, medieval Russia was less successful in warding off invaders from the East – including Mongols. One reason for the greater success of easterly moving invaders against Russian defenders, compared to Western ones, was, of course, the comparatively underdeveloped character of the Russian "state" in the former case. Another factor favoring the Mongols was their art of war and experience in waging fast moving, cavalry-centered battle over unprecedented distances. No other force, from medieval times to the present, has mastered strategic movement over terrain, endurance of weather and climate, and the ability to strike from several main directions under coordinated command-control, as the hordes of Genghis Khan and several of his immediate successors.

Geography defines the various media in which war takes place or might do so. Ancient and medieval warriors had to plan for war on land or at sea. Modern warfare takes place in, or is supported by, military activities on land, at sea, and in the air, space and cyberspace. The last of these, cyberspace, is not strictly speaking a "geographical" expression. But activities that the United States and other countries carry out in cyberspace have military import, perhaps decisively under propitious conditions of war. Therefore cyberspace is a fifth dimension for warfare, and perhaps a transgeographical one. But not

---

[3] Sun Tzu, *The Art of War*, translated and with an introduction by Samuel B. Griffith (Oxford: Oxford University Press, 1971), pp. 63–64 and passim.

so fast. Traditionalists would warn that armies, fleets and air forces cannot move through cyberspace. Killing zones require kinetic action that must take place in natural land, sea or airborne environments – and, perhaps within the present century, in space as well.

The traditionalists' caution is well intended, but cyber-enthusiasts would argue that the behavior space within which geography influences war has changed. The arrival of the information age means that geography has become mated to "infography" in compelling ways. Electrons that carry the vital information necessary for military forces to operate, including everything from C4ISTAR (command, control, communications, computers, intelligence, surveillance, targeting and reconnaissance) to logistics, do travel in cyberspace. An opponent who could disrupt this flow could conceivably shut down a military operation by turning its brain and central nervous system into sand. Indeed, in information age warfare, disruption is every bit as important as destruction. The ability to disrupt one state's art of war by denying it vital information in good time, compared to the performance of its opponent, is to bring the first state to the very threshold of defeat if not all the way over it. The preceding point was demonstrated by the Germans in 1940 against France, and by the U.S. and allied armed forces against Iraq in 1991.

Even with regard to computers and other aspects of military cyber operations, however, geography still matters. Hackers, like terrorists, must have safe havens from which to operate. Physical destruction of their locations may put them out of business temporarily or permanently. The challenge is to identify exactly who is hacking and from where. If hackers are located within a powerful state, such as China or Russia, kinetic responses are ruled out. Instead, cyber attackers may be vulnerable to cyber counterattacks. But even for that, we must know their GPS coordinates. Hackers, like other regular or irregular forces, require food, water, sleep and other necessities – and likely as not, they are on somebody's payroll. Nor is it necessary to counterattack them all – a few exemplary disappearances or lives lost might suffice to send object lessons. But, given the ease of internet-enabled cyber attacks and the numbers of persons attracted to this amusement, the defenders will always be outnumbered and forced to rely on considerable measures of self defense (e.g., firewalls, encryption) apart from the possibility of counterattack.

If the dimensions of future warfare include land, sea, air, space and cyberspace, nuclear weapons can be said to have a geographical aspect with respect to the first four, at least. Nuclear forces are based in certain locations, they are targeted on other locations, and they pose threats of intended and collateral damage that are considerable and unique. Nuclear weapons and nuclear warfare, if that threshold is ever crossed in the twenty-first century, are geographical, and therefore, geostrategic animals. From launchers based on land, at sea or airborne, nuclear weapons can be fired through the air and-or space, over land and sea, across distances that range from tens of kilometers to thousands of miles. The United States, with the world's largest number of deployed, long range nuclear weapons, can in theory hit any location on earth with some kind of nuclear charge, should it choose to do so. In practice, neither the United States nor any other nuclear armed state has an unlimited supply of ready weapons, so target plans must specify a finite menu of plausible options. These options must take into account many variables, including the exact locations of targets, topography (in the case of missile silos), weather (for enemy submarines tracking nuclear ballistic missile firing submarines), physical locations of vital military targets and infrastructure (bases, airfields, military manufacturing plants), and other location-dependent activities.

With respect to the construction of nuclear war plans for the purpose of deterrence, the sizes of notional countries of interest, in addition to other aspects of their geography, matter considerably. During the Cold War at various times, both the United States and Russia considered whether nuclear weapons would enable either side to deliver a knockout blow against the opponent's military forces and-or society. Planners faced the fact that America and Russia, geographically speaking, were actually continents or continental-sized civilizations as much as they were countries or nation-states. The scale of nuclear destruction required to destroy the Soviet or American armed forces during the high Cold War, under the most optimistic of assumptions about the performance of military systems, would also have rendered their societies, and those of their allies, as radioactive deserts. What drove this lesson home for the United States was the Cuban missile crisis, although U.S. and Soviet forces were smaller then than they would later become in the 1970s and 1980s. What drove the point home for the Soviet leadership was Chernobyl and its aftermath. Gorbachev, if not some of his more hidebound military advisors, got the point that the costs of "victory" in a U.S.–Soviet nuclear war would be unacceptable to the American and Soviet people and a crime against humanity – including socialist humanity.

It took the United States and the Soviet Union a long time to recognize that, beyond a certain point, the accumulation of more weapons did not lead to an equivalent increase in security. One reason for this is geography, both human and natural. There is a great deal of ruin in a nation, but only so much that can be meaningful to military accomplishment. Some U.S., and no doubt Soviet, plans for nuclear war required such redundancy in the destruction of military and other targets that the expenditure of ammunition and the firing of rockets would have represented a macabre triptych of Wagnerian opera more than a sensible war plan. On the other hand, it has to be said that planners did the best they could, under the exigent circumstances of their political tasking and in the face of the limitations of available technology – and geography.

For example, it soon became clear to U.S. and Soviet planners that firing transcontinental or transoceanic ballistic missiles over the North Pole toward targets in Siberia or North Dakota made more sense than it did to fire them across the Atlantic or Pacific. On the other hand, nuclear war might begin in Europe or Asia as a conventional war, escalating to nuclear first use. So some significant portion of American and allied NATO nuclear ordnance and delivery systems had to be reserved for use in European or other theater operations. The Soviet Union, on its side, had to be concerned with its eastern flank, facing a nuclear armed China and an American-protected Japan. "Theater" or "local" nuclear war planning required, for the Americans and for the Soviets, nuanced political and military guidance beyond the singular war plan for all cases and all out destruction.

Whether tasked for local or global thermonuclear war, American, allied NATO and Soviet military planners faced geographical and geostrategic speed bumps. In the early years of the nuclear age, for example, U.S. targeting of the Soviet Union was done largely by intuition and osmosis. Maps of the interior of the USSR available in the later 1940s and early 1950s were often based on World War II German intelligence sources or Soviet defectors' memories. Military overflights of Soviet territory by hostile aircraft were always dangerous, and it is a reasonable inference that more intelligence missions crashed and burned inside Russia than U.S. government officials have acknowledged. The Soviet shootdown of the U-2 flown by Francis Gary Powers created a large political embarrassment for the Eisenhower administration. It was not until the development of space reconnaissance satellites that the United States could establish with confidence the size and character of Soviet long range missile and aircraft deployments. And when it did, it created a political firestorm in the Kremlin, leading to Khrushchev's dangerous Cuban missile gambit as a short-cut to reversing the sudden appearance of American nuclear superiority.

Imaging satellites and other telltales, including satellites and other instruments for collecting signals intelligence, provided an abundance of data for U.S. and Soviet intelligence to pore over during the Cold War. However, geography still imposed limits. Overhead reconnaissance could not be conducted with equal vigilance over all locations of interest – there were simply not enough space borne or other platforms available. This meant, in the U.S. case at least, that interagency controversies over the tasking of photo-reconnaissance satellites resulted in the creation of an interdepartmental referee for the purpose of setting priorities. In addition, satellites and other above-ground observations could not determine some important parameters of interest to nuclear target planners. For example, the single shot kill probability of an American land or sea based strategic missile (ICBM or SLBM) against a Soviet missile silo would have depended upon the hardness of the missile silo, the kind of soil in which it was embedded, whether the attacking warheads were ground or air burst, and other factors. In addition, Soviet ICBMs were not randomly located within its state territory, but purposely distributed across the breadth of the USSR in order to minimize the likelihood of any successful U.S. first strike or, in the event of a Soviet first strike, U.S. counterforce retaliation.

If land based missiles posed targeting problems partly on account of geography and the limits of technology related to geography, submarine launched ballistic missiles were even more demanding. One could observe the locations of ballistic missile submarines (SSBNs) when in port or departing from port for patrol destinations. But once at sea, "boomers" defied detection by overhead reconnaissance and locating them, if possible at all, required sophisticated antisubmarine warfare forces (primarily state of the art attack submarines, or SSNs). From public sources, it appears that the U.S. boomers, at least, were virtually undetectable by hostile forces during the Cold War.

After the Cold War and the disintegration of the Soviet Union, U.S. ballistic missile submarines faced no peer competitor, and no ASW force capable of global operations stalked them. However, the boomers were potentially vulnerable when in port, during the Cold War and afterward, which is why a certain percentage of them were always kept at sea. Conversely, the Russian fleet ballistic missile submarine force fell into hard times with the collapse of the Soviet Union. Russia's economic problems in the 1990s impacted on defense spending, with the nuclear navy taking an especially hard hit. Plans to modernize the force were announced but not followed through. By the time Putin assumed the office of President in 2000, the sea based leg of the Russian nuclear "triad" was in serious trouble, relative to Russia's force of strategic land based missiles. During Putin's presidential term, the Russian economy improved over the preceding decade, and additional funds were devoted to military modernization. Russia has begun to deploy a new class (Borey) of ballistic missile submarines and is testing a new SLBM (Bulava) for eventual deployment with those boats. Russia recognizes that the SSBNs constitute a uniquely survivable deterrent that is required for even the appearance of nuclear-strategic parity with the United States, let alone the reality of survivable forces.

In the case of ballistic missile submarines, compared to land based forces, geography imposes challenges relative to reliable command and control. Submarines, like other maritime forces, must operate with a certain degree of tactical and operational flexibility, compared to forces that are land based. Admittedly modern communications deny captains at sea the autonomy that great "ships of the line" used to have in the heyday of sail. Still, it would be self defeating for submarine commanders not to have the wherewithal for timely and prompt decisions about maneuver and, if necessary, firing their weapons without prior clearance from regional or higher headquarters.

Submarines capable of firing nuclear armed ballistic or cruise missiles require much of the same tactical and operational flexibility as do those equipped with only conventional weapons. But no head of state or chief of the general staff wants to leave the decision for nuclear first use or retaliation, especially the decision to launch transoceanic ballistic missiles, to a tactical commander. Therefore the chain of command will impose certain constraints on nuclear capable SSBN and SSN commanders, including the need for redundancy in the clarification of launch orders, careful screening of personnel for reliability, and specialized message systems from higher headquarters to the submarines on patrol. In the case of the Soviet Union, its Cold War nuclear firing attack or ballistic missile submarines were undoubtedly staffed with political commissars whose approval was also required before launch orders could be authenticated and implemented.

Compared to land based forces and submarines, it might seem that strategic nuclear bomber forces appear to have fewer constraints imposed by geography or geostrategy. That impression is misleading. Bombers require basing, and the operation of bomber bases for planes capable of delivering nuclear weapons over transoceanic or transcontinental distances is actually quite complicated. The first constraint imposed by geography, combined with technology, is that bombers cannot themselves carry enough fuel for intercontinental missions. They therefore require aerial refueling, a technique that is challenging even for the United States. Few states have succeeded in mastering this skill, and none on the scale that the United States is able to do. The reason for this is that, during the Cold War and even afterward, no other state even attempted to field a long range bomber force of the size and complexity of the United States airborne fleet. Modernization of the U.S. bomber force has been sporadic over the decades of the Cold War and subsequently. Periods of high thrust in research, development and deployment have alternated with slow rolls. Nevertheless, old platforms endure: the redoubtable B-52 was still flying as a component of the U.S. nuclear strategic triad into the twenty-first century.

The command and control of bomber forces is also more complicated than it appears to be. This is, ironically, because bombers have one "advantage" over missiles: they can be sent aloft, flown partway toward their targets and then recalled if crisis conditions change. During much of the Cold War, a certain percentage of the U.S. strategic bomber force was always kept on ready or strip alert so that they could be scrambled out of harm's way in case of a Soviet first strike. If, however, a nuclear crisis were protracted, bombers would have had to return to base: they could not stay aloft indefinitely, or even as long as ballistic missile submarines could be kept dispersed and hidden at sea. Bombers could move from one location to another during a crisis as a signal to the other side of U.S. seriousness about the stakes. This use of bombers for political signaling happened more than once during the Cold War, including in contretemps with both the Soviet Union and China. Forward basing of nuclear capable bombers, say in Guam, did not commit the United States to irrevocable nuclear conflict, but it did create a greater capability for delivery of munitions of choice noted by potential adversaries and others.

Another geographical or geostrategic aspect of bombers was their slow rate of flying, compared to missiles. This attribute made missiles the preferred weapons for nuclear first strikes, once they were available in sufficient numbers, and bombers more appropriate for delayed as opposed to prompt missions. In addition, the technical capability for MIRVing long range ballistic missiles (placing multiple warheads on a single missile and allowing each warhead to be directed to a separate target) increased the destructive capability of U.S. and Soviet ballistic missiles by a significant factor – especially if either force was employed in a first strike. MIRV was also attractive to military planners and defense budget managers because it enabled more "bang for the buck": each single ballistic missile could delivery ordnance against as many as ten targets, as in the case of the largest Soviet intercontinental land based missile (the SS-18). As a result, MIRVed ballistic missiles took on the aspect of the "bad guys" that threatened deterrence and crisis stability, compared to the bomber "good guys" whose platforms were now almost certainly confined to retaliation after attack. In addition, mistaken launches of bombers could, as noted above, be recalled before Armageddon ensued: ballistic missiles lacked this option (some scientists proposed the addition of a "command destruct" option for ballistic missiles, but policy makers and military commanders found the temptation resistible).

Another asymmetry between Cold War bomber and missile forces, related to the overlap between geostrategy and technology, was the problem of defense against nuclear attack. As is well known, during the later 1940s and much of the 1950s, the United States and the Soviet Union anticipated that a third world war would involve large scale attacks by nuclear bomber forces. It was thought that air defenses could pose a credible threat of mitigation against bomber forces, protecting both military targets as well as civilian populations from mass destruction. However, once the intercontinental ballistic missile forces superseded long range bombers as weapons of choice for prompt attacks, this relationship between the prospective attacker and defender changed. Air defenses might have promised some meaningful attrition against bombers, but no antimissile defenses worthy of the name were available or on the drawing boards. Admittedly the Americans and Soviets both tried hard – both states fielded more than one generation of strategic antimissile defenses from the 1960s to the 1980s. But not even Ronald Reagan's dreams of a nationwide missile shield that would eventually make nuclear weapons "impotent and obsolete" stood a chance in the 1980s against the threat posed by Soviet ballistic missile attack. The Soviets, relative to the possibility of a U.S. nuclear missile attack, faced the same cul-de-sac.

Missile defenses – technology and geography

Eventually negotiations between the United States and the Soviet Union codified this counterintuitive relationship, between nuclear offenses and antinuclear defenses, into an arms control agreement (the ABM Treaty of 1972). It remained as one of the diplomatic benchmarks between the Cold War nuclear superpowers, along with the original Nuclear Test Ban Treaty and the various SALT and START agreements. For some members of the U.S. arms control community, the ABM Treaty was more than that. It symbolized deterrence stability based on mutual vulnerability of both American and Soviet societies to unacceptable retaliation. Mutual assured destruction or assured retaliation was, according to this school of thought, not a choice, but a condition, imposed by technology.

In fact, technology did favor the offense, but that did not necessarily argue for disinterest in antimissile defenses. After all, defenses could have been deployed and tasked with the protection of retaliatory forces: consistent with M.A.D. doctrine. The Nixon administration, seeking to begin deployment of limited missile defenses without jeopardizing talks with the Soviets about nuclear arms control, renamed the Sentinel ABM system as Safeguard and changed its primary mission from population protection to defense of ICBMs in the Midwestern United States. However, Sentinel-Safeguard was involved in public and Congressional controversy from the time of its first proposal by then Secretary of Defense Robert McNamara in the Johnson administration. Safeguard was finally deployed in 1975, but the House voted to deactivate the system almost immediately after it became operational. The SALT I and ABM signature agreements on offensive and defensive strategic arms limitations pushed antimissile defenses further onto the backburner of research and development, not to mention deployment, until the administration of Ronald Reagan.

One of the underappreciated points about missile defenses, relative to offenses, during the Cold War was that the supremacy of offenses was not only imposed by technology. It was also a creation of geography. The task of missile defenses was made impossible, not only because it was difficult to "hit a bullet with a bullet," but also because there was so much American or Soviet territory to protect. The attacker could strike with nuclear armed ballistic missiles against almost any portion of the defender's retaliatory forces, and the defender could strike back with equal or greater "unacceptable" damage against the forces, society and economy of the attacker. There was worse. In addition to the challenge of defending their respective state territories, the United States and the Soviet Union also had to defend their allies against nuclear first strikes or retaliation. This problem bedeviled NATO even without the complexity of missile defenses. How extensive was the "extended" deterrence provided by the U.S. nuclear umbrella? Could the non-nuclear state members of Cold War NATO rely on U.S. willingness to risk nuclear destruction of the American homeland in order to deter, or if necessary, retaliate against, a Soviet attack?

The problem was based on the antithesis between the requirement in NATO and American strategy for two necessary, but opposed, concepts: coupling, and firebreaks. Nuclear coupling implied persuading the Soviets that any use of nuclear weapons in Europe would involve automatic American involvement, and presumably, escalation to global nuclear war. On the other hand, firebreaks of two sorts were argued for by some theorists and policy makers, usually Americans. The first firebreak was between conventional and nuclear war; the second, between theater nuclear war in Europe and so-called strategic or global nuclear war between the United States and the Soviet Union. NATO lived out its Cold War existence with these two objectives poised in Kantian juxtaposition, and carrying very different implications for NATO policy and strategy.

Those who emphasized nuclear coupling in NATO strategy deliberations emphasized that only the expectation of a seamless web of escalation, from nuclear first use to strikes against the American and Soviet homelands, would serve as a reliable deterrent to *any* war in Europe. The counterpoint was that overemphasis on coupling would prevent escalation control and conflict termination short of mutual societal destruction and, perhaps, global ecocide. A "man from Mars" might have agreed that nuclear coupling was less likely to fail as a deterrent, but, if it did, it was more likely to escape political control. The same observer might have agreed that firebreaks were more likely to allow for escalation control once deterrence had failed (either conventional or nuclear, or both), but an escalation ladder with prominent firebreaks might encourage brinkmanship and contribute to deterrence failure.

Both the emphasis on coupling and the argument for firebreaks in NATO doctrine and strategy could be based on some abstract logic. But nuclear abstraction in this case crashed into geographical reality. A "limited" war from the American perspective could be, and almost certainly would be if nuclear weapons were employed, a total war for Europeans. In addition, U.S. nuclear weapons and troops forward deployed in Europe would create an immediate fear of escalation to global war, even if the initial Soviet attack used only conventional weapons. Given the Soviets' geographical proximity to NATO'S European military assets and vital centers, as well as the Soviets' superior numbers of ground forces deployed in Europe, a NATO decision to remain below the nuclear threshold was tantamount to an acceptance of at least partial defeat.

Had the United States perfected a national missile defense for the American homeland in the 1980s, the deployment of such a system would have intersected with the firebreaks-coupling dilemma in the U.S.–NATO relationship. Would the missile defenses include NATO Europe, either by means of technology transfer or by extending the reach of American interceptors, sensors and command-control systems? The reaction of the Soviet Union to a plan for U.S. BMD protection for NATO Europe in the 1980s would have been less than inspiring of détente. In one stroke, Russia's nuclear deterrent against the United States would have been nullified, along with its option to use coercive diplomacy against Western Europe during any nuclear or other crisis. Under these conditions, Soviet leaders would either have had to accept de facto defeat in the Cold War or employ offensive countermeasures to U.S. extended defenses for Europe – perhaps reigniting a hotter phase of the Cold War that would otherwise have petered out.

Counterfactuals should never be followed into the wilderness of mirrors, so the "what ifs" of BMD counterfactuals have no tangible historical value. They only serve here for the purpose of demonstrating that one cannot get away from geography, either terrestrial or celestial, even if the most optimistic assumptions about technology triumphalism are made. As in the past, any future U.S. or other national missile defense system will be protecting some designated national and-or allied territories. These territories will presumably be the repositories of sovereign governments that claim a monopoly of legitimate force throughout their breadth and depth – including, for military and defense purposes, their national airspace and littoral waters. These territorial and national jurisdictions have symbolic and cultural importance: people will fight and die for them, regardless of the "rationality" of the reason for fighting. The war between the United Kingdom and Argentina over the Falklands-Malvinas is merely one example of the obduracy of nationalism attached to territory – in this case, territory about as remote, forbidding and inhospitable to normal creature comforts as travelers or military operations can abide.

The idea that space based missile defenses can supersede geography or geostrategy is mistaken, not only with respect to the past, but also with regard to the future of missile defenses or other advanced technologies. The information age has given rise to an informative, but dangerously seductive, vocabulary of terms that imply the irrelevance of all but cyber competencies in the operation of armed forces. But information technology cannot enable U.S. or other military forces with superior fighting power unless technology innovation is coordinated with changes in military organization, doctrine, and training. Armies, fleets and air forces must still deploy, and prepare to fight, in their respective environments. In addition, U.S. military doctrine requires "jointness" across arms of service for the planning and conduct of military operations by regional or functional combatant commands (as in Central Command, Pacific Command, and so forth).

The same realities will influence future decisions about U.S. or other missile defense deployments. Whom to protect, and against what, are the political decisions and threat assessments that will drive decision makers to choose among available technologies and weapons systems. These threat assessments will be influenced by geography and by assumptions about geostrategy. They will, for example, identify particular countries with specific locations, either as prospective ballistic missile attackers, or as territories to be defended against such attacks. For example, should North Korean nuclear and missile capabilities require that the United States provide missile defenses for Japan, that Japan establish its own nuclear weapons capability, or that some regional BMD system be deployed to protect both South Korea and Japan? Currently the United States is limiting its response to the first option, by working with Japan to deploy sea based ballistic missile defenses.

The Obama administration announced in September, 2009 that it would scrap the Bush plan for missile defenses in Europe. Instead, the United States would deploy a reconfigured system that was focused more on intercepting Iranian short or medium range missiles, as opposed to the longer range ballistic missiles that Iran might eventually develop.[4] According to U.S. Secretary of Defense Robert Gates, the new plan would actually deploy missile defenses against Iranian threats seven years ahead of the Bush plan, and some land based interceptor missiles would be deployed in Europe, possibly in Poland.[5] Anticipating some negative reactions to the plan from Europe and the American home front, Gates added that the new BMD configuration "provides a better missile defense capability" for Europe and

---

[4] Peter Baker, "White House to Scrap Bush's Approach to Missile Shield," New York Times, September 17, 2009, http://www.nytimes.com/2009/09/18/world/europe/18shield.html. See also: Michael D. Shear and Ann Scott Tyson, "Obama Shifts Focus of Missile Shield," Washington Post, September 18, 2009, http://www.washingtonpost.com/wp-dyn/content/article/2009/09/17/AR2009091700639.html.

[5] Baker, "White House to Scrap Bush's Approach to Missile Shield." For a favorable expert assessment of the Obama missile defense plan, see Hans Binnendijk, "A Sensible Decision: A Wider Protective Umbrella," Washington Times, September 30, 2009, in Johnson's Russia List 2009 – #181, September 30, 2009, davidjohnson@starpower.net. Continuing Russian doubts are noted in "Russia Still Suspicious of U.S. Missile Defense Plans," Reuters, September 29, 2009, in Johnson's Russia List 2009 – #181, September 30, 2009, davidjohnson@starpower.net.

U.S. forces located there than the program he had recommended almost three years earlier (as President George W. Bush's Secretary of Defense).[6]

The focal launch platform for the Obama missile defense plan would be the Standard SM-3 interceptor missile, currently deployed on U.S. Navy ships and eventually on land as well.[7] The first phase of the plan would be completed by 2011 with the deployment of sea based interceptors. The second phase, becoming operational around 2015, would deploy upgraded SM-3s in land based sites in Southern and Eastern Europe.[8] This revised plan was based on U.S. intelligence assessments that Iran's progress in the development and deployment of short and medium range missiles has speeded up, whereas its progress on longer range missiles has slowed down.[9] U.S. officials also indicated that another reason for the strategy and technology shift was the desire to get missile defenses into position sooner and closer to Israel. More readily available defenses against short term threats to Israel could help persuade Tel Aviv to withhold any preemptive or preventive attacks against Iran's nuclear program. As U.S. Secretary of Defense Robert Gates noted, "We hope that it will reassure them [Israel] that perhaps there's a little more time here."[10] Thus, the new U.S. missile defense plan was intended to be in tune with American goals for regional deterrence and crisis stability as well as for nonproliferation.

Remapping strategy: East and West

However, the inability of the United States and other nuclear negotiators with North Korea (South Korea, China, Russia, and Japan) to rewind history and achieve the disarmament of North Korea's nuclear capabilities may cause both Japan and South Korea to rethink their options. Japan's extensive civil nuclear power industry incorporates the nuclear fuel cycle necessary for the manufacture of bomb grade materials within months, should the government take a decision in favor of doing so. China's reaction to a decision by Japan for

---

[6] Baker, "White House to Scrap Bush's Approach to Missile Shield."

[7] Mission and technical information pertinent to the sea based SM-3 and related platforms appears in U.S. Department of Defense, Missile Defense Agency, *Fact Sheet: Aegis Ballistic Missile Defense* (Washington, D.C.: Missile Defense Agency, April 2009). See also: "Ticonderoga Class Guided-Missile Cruisers, USA," *Naval-Technology.com*, http://www.naval-technology.com/projects/burke/, undated, downloaded July 27, 2009, and *RIM-161 SM-3* (Aegis Ballistic Missile Defense), www.globalsecurity.org, downloaded October 22, 2009.

[8] Robert M. Gates, "A Better Missile Defense for a Safer Europe," *New York Times*, September 19, 2009, http://www.nytimes.com/2009/09/20/opinion/20gates.html.

[9] Baker, "White House to Scrap Bush's Approach to Missile Shield."

[10] Ibid.

nuclear weapons status would not be favorable, and an acceleration of its own nuclear weapons capabilities, including land and sea based ballistic missiles, would be among Beijing's options.

Here we see one of the reasons why nuclear proliferation in Asia might be more dangerous than in Europe. It is sometimes supposed that cultural or other domestic political reasons make nuclear arms races more dangerous outside of Europe, compared to the situation that obtained between the U.S. and Russia during the Cold War. Thus, for example, Indians and Pakistanis might be driven into war, including first and retaliatory uses of nuclear weapons, for reasons based on cultural antipathy, including ethno-national or religious hatred. The same fears have been expressed about a possible nuclear arms race in the Middle East: not only between Israelis and Iranians, for example, but also between Iranian Shi'ites and leading Sunni Arab state powers, including Saudi Arabia and Egypt. These concerns about states being driven beyond the brink by motives other than European-derived or Eurocentric political ideologies are certainly valid. States in the Middle East or Asia are not going to war over the distinction between communism and capitalism.

On the other hand, it is also the geography of Asia that makes the possibility of nuclear weapons spread so immediately threatening to international stability. Nuclear Asia is a high octane concentration of powerful and combustible appetites for growing international influence and respect within a very small geopolitical space. Stand in the middle of Urumqi in western China and draw a circumferential line from the west, through India and Pakistan, to the north, across southern Russia, southeast into China and North Korea, and across the Sea of Japan to the Japanese home islands. A traveler who flew across this route would have crossed over five acknowledged or self declared nuclear weapons states and a sixth (Japan) that was a turnkey away from joining the ranks. Worse is that some of these states with embedded memories of historical grievances are contiguous to one another: India and Pakistan, having fought three wars and a number of skirmishes during the Cold War; and North and South Korea, whose war beginning in 1950 has never been officially terminated by a final peace settlement (an armistice stopped the fighting in 1953).

So thanks to geography, combined with disruptive technologies of the industrial age (missiles and weapons of mass destruction), the revised strategic map of Asia, including those parts of the Middle East within missile reach of Asia, threatens the durability of the nuclear nonproliferation regime and raises the probability of nuclear first use in the twenty-first century.[11] However, nuclear Asia is not isolated from the rest of the world – thanks, in part, to geography,

---

[11] For an expansion of this point, see Paul Bracken, *Fire in the East: The Rise of Asian Military Power and the Second Nuclear Age* (New York: Harper Collins, 1999), esp. pp. 95–124.

combined with modern technology for military and commercial uses. Russia sprawls over eleven time zones, and in fact, most of Russia's state territory lies in Asia. Many of Russia's most important natural resources are found in Asiatic, not European, Russia. The low population density of Russia's Far East invites unofficial migration across Russia's extensive border with China.

Facing west, Russia's economic and political integration with Western Europe is prerequisite to the continuation of Europe's postmodern, debellicized status in the early twenty-first century. But a stable Russia available for partnership with NATO or with the G-8 assumes a Russia whose economy can continue to grow and whose democratic institutions will provide for a favorable business climate under the rule of law. The evolution of Russia in this westernized direction, and away from its more isolationist-narcissist proclivities of an imperial past (sometimes called "slavophilia" but this term has its limitations as an opposed term for westernization, since it implies that Russian nativist, antiintegrationist tendencies are based on race or nationality), cannot be taken for granted.[12] Russia will insist upon a certain security perimeter in its "near abroad" including former Soviet dominated parts of Europe.

NATO's expansion since the end of the Cold War troubles Russia, but not because Russia expects or fears a military invasion by NATO (although this hobgoblin is useful to Russian military atavists who are resisting reform of the Russian armed forces for other reasons). NATO enlargement bothers Russia for the same reason that U.S. missile defense deployments in Poland and the Czech Republic do – geography, and their implications for geostrategy. During the Cold War, Soviet conventional military superiority in Europe was offset by U.S. and allied NATO nuclear deterrence. In current, post-Cold War Europe, American and allied NATO conventional military forces are a generation ahead of Russia's in the requisites of information-based warfare: especially in capabilities for seeing and comprehending the battlespace; for sharing information within and across theaters of operations; and for delivering long range, precision strikes against a variety of fixed and mobile targets. U.S. supremacy in these "C4ISTAR" dimensions of modern warfare (command, control, communications, computers, intelligence, surveillance, targeting and reconnaissance) encourage Russians, including Russian military planners and political leaders, to rely on the threat of nuclear first use more than hitherto to deter any outbreak of conventional war in Europe threatening to Russia.

---

[12] A concise discussion of this issue appears in Anders Aslund and Andrew Kuchins, *The Russia Balance Sheet* (Washington, D.C.: Peterson Institute for International Economics and Center for Strategic and International Studies, April 2009), pp. 14–15. The many faces and causes of Russian empire are discussed in Philip Longworth, *Russia: The Once and Future Empire from Pre-History to Putin* (New York: St. Martin's Press, 2005).

In part, these technology driven Russian fears are oxymoronic. Neither the United States nor NATO has any intention of invading or attacking Russia. To the contrary, the U.S. administration under President Barack Obama has indicated its desire to "reset" relations with Russia in a favorable direction, including in security and arms control matters. Unfortunately this shift in U.S. approaches is not the end of the matter, and Russia's fears are not based mainly on technology. They are based mostly on history, including Russia's understanding of its history from a post-Cold War perspective, and geography. Post-Cold War Russia, especially under Putin, saw itself as under siege from NATO enlargement, from rebels in Chechnya and elsewhere in the Caucasus, from demands for pluralist democracy inconsistent with Russian traditions, and from a potential security meltdown near its borders as a result of political revolutions in Georgia and Ukraine. Admittedly these are blinkered perceptions of Russian *siloviki* and others in Putin's entourage, but those perceptions rest on a wider base of Russian experience with invasions from the west by Sweden, France and Germany – all during "modern" European history.

Currently Russia's leadership is not fearful of a military invasion, but of the creep of democratic institutions to the doorstep of Russia or over the threshold. Both Medvedev and Putin favor guided democracy with a dash of unofficially approved corruption and elitism, but not a return to communism. They and their fellow crony capitalists, including some who also hold positions of political influence or government office, would be at risk without a manipulated media sector, controls over the activities of political opposition, and a top-down federal government bureaucracy, especially in the security sector.

The Rose and Orange revolutions unsettled the Kremlin because of the lethal combination of geography and democracy of the sort that Moscow does not want to import. Georgia and Ukraine are both adjacent to Russia, Ukraine was historically a part of the Russian and Soviet empires, and Georgia was a former Soviet republic. The incorporation of either Georgia or Ukraine, especially the latter, into NATO places that alliance's military-strategic umbrella over a democratic dagger pointed at Russia's heart. In addition, and from Russia's geostrategic perspective, NATO membership for Ukraine also pushes NATO's border that much further into Russia's historic heartland, creating a contiguous security risk. NATO membership for Georgia, less immediately threatening but still opposed by Russia, would in Russia's view open the door to insurgency and terrorism from jihadists and others crossing from sanctuaries in Georgia into southern Russia.

NATO has some dilemmas too, with respect to the plans for enlargement to include eventual membership for Georgia or Ukraine. NATO is a military alliance with a guaranty that an attack on any member of that alliance is an attack on all, calling forth a collective military response. Think what this might have meant in August, 2008 during Russia's war against Georgia, had Georgia already been accepted into NATO membership or even into membership-in-waiting via the Partnership for Peace. Even for the United States with its global military reach, geography imposes stark choices. Would the United States have been able in the autumn of 2008 to undertake a military campaign in the Caucasus, given its existing (and exhausting of manpower) commitments to Iraq and Afghanistan? Even after American and NATO military commitments in Iraq and Afghanistan have evolved into a lesser drain on their resources, deployment of ground troops and supporting elements into the Caucasus by NATO over Russian resistance could create a literature to rival Tolstoy's classic accounts of Caucasian warfare. Truth be told, the United States is unlikely to be withdrawing all its forces in Afghanistan any time soon. In fact, the Obama administration has already redeployed some forces from Iraq to Afghanistan in order to increase U.S. and NATO fighting power there, and the former commander of United States and allied NATO forces in Afghanistan, Gen. Stanley McChrystal, urged additional troop deployments in fall, 2009 as a necessary means of avoiding political and military defeat.

If the image of American and NATO military actions in the Caucasus vs. Russia creates enduring images worthy of Jerry Bruckheimer, then imagine a U.S.–NATO military operation in NATO member Ukraine, opposed by Russia. Recall that NATO is an alliance with nuclear weapons, and that Russian military doctrine calls for nuclear first use (presumably, with short range or tactical nuclear weapons) in the case of conventional wars near, or within, Russia's borders, and with the potential to do strategic harm to Russia.[13] However, for the sake of argument, let us suppose that somehow military clashes between NATO and Russia avoid the use of nuclear weapons by either side, and a truce is eventually signed that permits continued membership by Ukraine in NATO.

NATO now has the task of nation rebuilding in Ukraine, dealing with antiregime partisan activity originating in Russia and supported by indigenous pro-Russian elements (viz., Abkhazia and South Ossetia in Georgia). Count the numbers of Ukrainians that will require adult supervision by security services of various kinds, both military and police forces contributed by NATO and by others, and including Russians according to the war termination agreement. As this postconflict security and stability operation seeks to reconstruct war torn Ukraine and reconstitute a democratic regime there, publics and legislatures

---

[13] Prospective revisions in Russian military doctrine might strengthen the propensity for nuclear first use in local as well as regional wars. See: David Nowak, "Russia to allow pre-emptive nukes," *Associated Press*, October 14, 2009, and "Russia's Message on Reshaping Its Nuclear Doctrine," Stratfor.com, October 15, 2009, both in *Johnson's Russia List* 2009 – #190, October 15, 2009, davidjohnson@starpower.net.

in NATO become restive as casualties mount, expenses increase and doubts about the extent of NATO's security commitments rise. This scenario for a postconflict Ukraine is not at all as fanciful as it might appear. Historians have tracked the singularly brutal partisan warfare on the eastern front in World War II, and especially in Ukraine, where Ukrainian nationalists fought against the Germans, against the Soviets, and against one another. They continued fighting after the end of the war against the Soviets.[14]

Even if the transconflict or postconflict situation is more stable than this scenario admits, nuclear weapons remain in the background of any conflict termination in Ukraine or, for that matter, in Georgia as the site for a war between NATO and Russian military elements. These nuclear weapons will not "win" a war for Russia or for NATO because they are not weapons for prevailing in combat at an acceptable cost, but for deterrence against further military escalation by either side. Nuclear weapons impose some discipline over threats and combat actions as between the two sides, but the conventional battle between pro-NATO Georgians or Ukrainians and their domestic political opponents, supported by NATO and Russian outsiders, has an uncertain gravitational force. A massive Russian conventional military intervention in Ukraine or Georgia might force NATO to up the ante by attacking the bases from which Russian forces embarked, either in the North Caucasus military district or in European Russia.

In either theater of operations, in addition to the other difficulties in containing an outbreak of war, geography runs out of rubber room very quickly. Even Ukraine, much larger than Georgia, could find itself in the position of Poland in August, 1939 when German and Soviet forces conducted invasions and occupations from separate directions (pursuant to the Molotov–Ribbentrop pact). NATO forces could hug the western part of Ukraine where they would have relatively more sympathy, while Russian forces attempted to establish a rump Russian-dominated regime in the eastern part of the country. Both NATO and Russia could then be faced with nationalist uprisings against all outsiders, requiring cooperation against geographically separated, but strategically coordinated, insurgencies and campaigns of terror. Meanwhile, events in "old" Europe would be taking their own toll on NATO's determination to hold the line in Ukraine, and even alliance members from Rumsfeld's "new" Europe would begin to sweat profusely over the possibility of military escalation. Poland and the Czech Republic, for example, would worry that Russia might squeeze harder against their willingness to base missile defenses made in the United States on their state territories. Russia might even remind leaders of both countries that, in the event of a wider war that expands beyond Ukrainian

---

[14] Richard Overy, *Russia's War* (New York: Penguin Books, 1998), pp. 125–153.

territory, BMD deployed in Poland and in the Czech Republic would invite preemptive attack (on the grounds that they would contribute to nullification of Russia's nuclear deterrent at the very time that it was most needed for coercive diplomacy by Russia, or worse). This last point might be less arguable by Russia against the Obama missile defense plan, compared to the George W. Bush program, although the Obama plan does envision some land based antimissile deployments and eventual engagement of intermediate, and even intercontinental range, offensive missiles.

Asymmetry agonistes and command-control

Having read this far, some unrepentant neocons and other imperialists-light are reacting as follows. The authors underestimate American military capabilities for control of the strategic "commons" against all comers: including the seas, air and space media. In addition, the United States can conduct conventional deep strikes against targets with relative impunity and rapidly deploy forces into almost any contested area on short notice. During conventional deep strikes or rapid deployments, U.S. and allied NATO military operations will be supported by one of a kind logistics and information superiority. All of this is true, and it fuels a sense of military entitlement that is perhaps overdrawn at the bank. Military superiority in "system of systems" for advanced technology conventional warfare is not necessarily transferable into all situations of armed suasion.

Asymmetrical responses to apparent U.S. superiority in conventional warfare include insurgency and terrorism (at the substate level) and weapons of mass destruction carried by ballistic or other delivery systems (at the state or interstate level). A2AD (antiaccess and area denial) means of asymmetrical warfare will be used by future U.S. and NATO opponents for excluding or deterring military interventions in their respective backyards. These means of asymmetrical warfare might include, for terrorists or insurgents who are dedicated enough, suicides and homicides en masse. The detonation of an ammunition dump covered with radiological debris, biotoxins, or other unsavory materials in an urban area could complicate insertion of U.S. or other foreign troops into hostile zones. So, too, could the use of ballistic missiles of various ranges in order to deliver biological or chemical charges against the logistics and communications tail of U.S. or allied expeditionary troops engaged in activities as politically uncontroversial as humanitarian rescue (but for which side, and with what rules of engagement) to forces set upon the more politically contentious mission of regime change.

The list of asymmetrical options against deploying U.S. or other forces by regional adversaries of all stripes, including unconventional ones, admits of endless possibilities and typologies. All of these instruments of destruction and insurrection must, however, take place within a geographically bounded and defined security space. Even if one stretches the imagination to include a global security space or a regional one, geography underlies any operational definition of security and constrains the military actions taken on behalf of that concept. The Mongols without a Central Eurasian steppe on which to perfect their military art would have been just another tribe, and not a military phenomenon. A United States of America that halted its westward expansion at the Mississippi would never have attained its present status as a world power. An England that did not depend on transoceanic commerce for its economic growth and development, indeed, for its very survival, would not have been motivated to develop a world class navy, nor to enforce the pax Brittanica over an unprecedented global landscape. But geography is also a destiny that lures states to ruin, including Britain and Russia in Afghanistan and the Americans in Vietnam.

In the latter case, of the Americans in Vietnam, geography placed U.S. President Lyndon B. Johnson in an untenable political and military position. The U.S, could not win the ground war by fighting only in South Vietnam: invasion of North Vietnam was the only way to disconnect the insurgency in the south from external, and decisive, support. On the other hand, in the midst of the Cold War, expanding the American ground war into North Vietnam risked drawing in both China and Russia, especially the former, as official belligerents. And China was now a nuclear power, apart from its capacity for sending endless numbers of "volunteers" into the fight in North Vietnam (as it had sent hundreds of thousands into the fight on the side of North Korea during the Korean War). Escalation of the ground war into North Vietnam, even if it had been feasible in terms of U.S. domestic public opinion and available military manpower, would have occurred under the shadow of Chinese and Russian nuclear weapons. Those nuclear weapons, mated to ballistic missiles of various ranges, would not have had to be fired by Russian or Chinese operators in order to send a message of resolve and deterrence. They would merely have had to be positioned close to the immediate theater of operations in North Vietnam to reduce the risk of American nuclear escalation in the face of conventional military defeat.

Geography, in other words, imposed restraint not only on the conventional war in Vietnam, but also on the ability of the United States to make credible threats of nuclear escalation in a wider war. Russian and Chinese short or medium range ballistic missiles and tactical aircraft could have credibly threatened to impose local and regional nuclear costs for any U.S. tactical use of nukes against North Vietnamese or allied forces. But it was not only the tactical capability of these Soviet and Chinese forces over a specific geographical area that, by itself, gave them their potentially dissuasive power. In addition, it was their connectedness to a larger and potentially more destructive chain of escalation that might, if it escaped reliable control by either side, erupt into war in Europe or, in the worst case, a global war. The preceding point should not be misunderstood. Escalation of the American war in Vietnam into a regional or larger nuclear conflict, after a U.S. ground invasion of North Vietnam in order to depose the regime, was neither inevitable nor logically compelling as an option for policy makers. The invasion embodied unacceptable risks because it occupied the never-never land of risk: neither inevitable, nor impossible. It was a manipulation of the risk of uncontrolled escalation, under conditions of not only uncertainty (outcomes are known, but reliable probabilities are not attached to them) but also indeterminacy for relevant outcomes.

In addition to its uncertain or indeterminate, but important, relationship to nuclear escalation, geography also influences the aspect of command and control over nuclear forces. This point has already occurred in another context, but the present discussion is specific to the issue of attacks on political and military command and control systems during wartime. In the logic of conventional war, it follows that wartime leadership, both political and military, offer legitimate targets for hostile ordnance. Although assassination as such is often not condoned, depending on the perspective taken by governments relative to international law, it has been accepted that heads of state who are wartime leaders, and their principal advisors, might be subject to enemy attack, including killing or capture. Thus U.S. President Franklin Roosevelt and British Prime Minister Winston Churchill arranged a clandestine meeting in August, 1941 at Placentia Bay off the coast of Newfoundland – press accounts were blacked out until after the fact. U.S. and British fears were not of media overexposure, but of German torpedoes.

It has also followed, at least in modern times, that wartime military commanders are also subject to being killed in combat or capture. Japan's Admiral Yamamoto Isoroku, architect of the operations plan for the attack on Pearl Harbor in December, 1941, was specifically targeted and shot down over the Solomon Islands several years later by U.S. naval aircraft. And the U.S. government ordered General Douglas MacArthur to evacuate the Philippines for fear of his high profile capture (or worse) at the hands of invading Japanese forces. What the Soviets would have done with Adolf Hitler, had they been able to capture him alive, is uncertain – but if Hitler, instead of taking a decision for suicide, had ordered his bodyguards to fight to the death with him from the Fuhrerbunker, Russian infantry would have obliged him.

On the other hand, there are circumstances even in conventional war when it makes some sense to preserve the lives of surviving enemy military commanders or national leaders. The United States eventually decided that it was worthwhile to spare the life of Japanese emperor Hirohito and to leave him on the throne as a unifying cultural symbol in order to expedite Japan's World War II surrender and postwar U.S. occupation. This decision came rather late in the day, however: during earlier American air attacks on Tokyo, bombs had fallen more than once within the grounds of the Imperial Palace. In addition, heads of state and government are aware that, in time of total war, military defeat may place their lives and liberties at imminent risk. As German forces pushed into the western Soviet Union in Operation Barbarossa, early Soviet losses caused Stalin to consider relocating the entire government from Moscow to Kuybyshev. Although a great deal of wartime production was moved from western Russia eastward, the Soviet leader and his entourage remained in the Kremlin.

Nuclear war, as it might have been fought between the United States and the Soviet Union during the last two decades of the Cold War, presented the possibility that strategy might require a bipolar approach to the survival of political leaders and military commanders. That is: however an exchange of nuclear weapons might have begun, political leaderships on both sides would have some incentive not to permit escalation to continue on automatic pilot until arsenals were exhausted and entire populations and habitats were destroyed. For termination of any large scale nuclear war, some surviving and authoritative political leaders on both sides would have to negotiate a "cease fire" at least temporarily. This might be no easy matter to arrange, since early attacks could have destroyed some command-control facilities and incapacitated or killed a number of political leaders and-or military commanders.

In the United States, the issue of command and control survivability also affected planning for nuclear war and, therefore, expectations about the impact of command and control arrangements for deterrence. Nuclear forces had to be made proof against accidental or inadvertent first use or first strike; equally important, the same forces had to be responsive to duly authorized commands. The U.S. command and control system was designed over several decades to accomplish both of these objectives, and the accomplishment, such as it was, was both made possible, and made difficult, by geography. Geography dictated that surviving force commanders and political leaders could only be located in so many secure facilities, connected to others in the military or political chain of command. There were limits to how far the U.S. Presidential "center" (the President and his or her immediate retinue, including key political and military advisors and secure equipment) could be dispersed if the President was to retain the function of commander in chief.

President George W. Bush discovered this fact during his peregrination around the country in the immediate aftermath of 9/11. Eventually he insisted on a return to Washington, D.C. where he could reestablish a visible command presence to reassure Americans.

In order to preclude any enemy "decapitation" strike against the President that might otherwise disable the government from appropriate response, the United States has in place both a political line of succession and a military delegation of authority. The death or incapacitation of the American President automatically passes the office of President to the Vice President, the Speaker of the U.S. House of Representatives, the President Pro-Tem of the U.S. Senate, and then to the heads of Cabinet departments in order of their creation. The military delegation of command authority, on the other hand, runs from the President to the Secretary of Defense, "through" the Joint Chiefs of Staff and directly to the regional or functional military commanders. In theory, all of this was intended, not so much for the survival of an actual nuclear war, but to raise the uncertainty level of any prospective attacker. Further confusing to presumed evil doers plotting nuclear attacks was the possibility that, under certain circumstances that were veiled in secrecy even more than most, surviving air force and navy commanders and-or their subordinates could be empowered and enabled to continue nuclear strikes beyond the early phases of a nuclear war. Academic theorists and think tank analysts proposed at various times that the United States or the Soviet Union might be able to fight and survive a protracted nuclear war, although the idea lent itself to a high giggle factor among military insiders and night club comedians.

If we recall that all of this labor intensive preparation for command survivability was for the purpose of preventing war, rather than actually having to fight a war, this nuclear command-control survivability issue sounds somewhat less macabre. However, even those who would have pressed the "folly" button at the thought of a controlled or protracted nuclear war, involving tens of thousands of U.S. and Soviet nuclear warheads, would have to admit that the possible outbreak of a conventional war in Europe, amid predeployed American and Soviet tactical nuclear weapons, had an immediacy that departed from risibility. NATO had deployed nuclear weapons for use with operational field armies and air forces within the European theater of operations. In time of crisis, some of these nuclear charges would have to have been moved from peacetime storage sites to other locations, and some might even be mated to delivery systems such as missiles or bombers to avoid conventional or nuclear preemption.

Soviet intelligence, in its turn, would have observed NATO movements of nuclear munitions from stored to usable positions, and this information would have ratcheted up Kremlin sensitivities to other signs of NATO readiness for war. Soviet countermeasures to NATO readiness measures would then be followed by NATO counter-countermeasures, and so on, until some NATO or Soviet field commander raised the issue of nuclear release authority for forces under his command. Something like this action–reaction pattern of expectations allegedly happened during a NATO command post exercise in 1983 designated Able Archer.[15] The exercise was intended to simulate the decisions and actions that would be required for NATO nuclear release during a crisis. Apparently some Soviet intelligence sources felt the exercise was too realistic and passed warnings to higher authorities of a possible U.S. or NATO nuclear first strike. These misperceptions or exaggerations by fearful Soviet intelligence collectors were part of a mindset encouraged by some KGB and military leaders who had previously set in motion a collection operation against NATO called RYAN, an acronym for a nuclear surprise attack.

Had a nuclear war begun in Europe with the use of NATO or Soviet tactical nuclear weapons, the influence of geography on the course of events, including the likelihood of escalation to a wider war, would have been decisive. Within a small geographical space compared to the Soviet Union or North America, two opposed coalitions had commingled conventional and nuclear weapons, forces trained and tasked for both kinds of warfare, and war plans and exercises supposedly accommodating a transition from one kind of warfare to another. In addition, Soviet war plans apparently included preparations for prompt wartime attacks on both military and political command-control targets, including NATO political leaders perhaps targeted for assassination by Soviet special forces. Even prior to the introduction of nuclear weapons into this melee, command and control over the combatant forces of NATO deployed in Germany or Belgium would have been subjected to unprecedented strain, further increasing the possibility of military defeat and-or nuclear escalation out of desperation. However, the first use of nuclear weapons by NATO in West Germany and the probable response from Moscow would in short order have weakened knees in Bonn as well as in Brussels, opening intramural political debates about whether further resistance was militarily pointless or even humane.

Further to the issue of geography, the French, in and out of NATO from the mid-1960s on, were a nuclear fudge factor. Their nuclear *force de frappe* (later, *force de dissuasion*) was tasked both for the eventuality of supporting NATO and for the necessity of French national self defense – but primarily for the latter. Its notional targeting for "all azimuths" was a geographical expression of

---

[15] See Christopher Andrew and Oleg Gordievsky, *KGB: The Inside Story Of Its Foreign Operations from Lenin to Gorbachev* (New York: Harper Perrenial, 1991), pp. 583–600 for pertinent background on Able Archer and the worldwide Soviet intelligence operation RYAN which was part of the pertinent background.

a military and political ambiguity deliberately created for its strategic effect. If a Soviet conventional or conventional and nuclear military campaign advanced to the point at which France feared for its vital interests, it reserved the option to launch preemptive nuclear strikes, either against Soviet forces already on NATO (but not necessarily French) territory, or, under greater exigency, against targets located in the Soviet Union (read: Moscow). This threat to "tear an arm off" the Soviet Union implied that France alone, whatever NATO chose to do, could create for Moscow costs disproportionate to the assumed gains from any offensive threatening to France. To reinforce this message, France deployed not only nuclear weapons for delivery by aircraft and land based missiles, but also ballistic missile submarines presumably invulnerable to any Soviet first strike. Presumably French films would be air dropped only as a last resort.

Soviet plans for command-control survivability in the event of a nuclear war were neither more credible nor less complicated than those of the United States and NATO. Although Soviet military writings extolled the virtues of civil defense and strategic antimissile and antiair defenses in wartime, actual Soviet preparations for nuclear command-control survivability were bounded by the limits of science, politics and geography. Secure redoubts and bunkers could be constructed for the party, government, military and security elites, but they might inherit a country pushed backward into the economic and social conditions of the Russian "time of troubles" in the early seventeenth century – or worse, with radioactivity added in. Geography dictated that strategic defenses even more advanced than those of the Cold War would still have had to be tasked preferentially – in terms of protecting retaliatory forces, as opposed to populations, and among populations, establishing which cities receive preferential protection in case of limited resources. The decision to deploy the antimissile systems that were available for the defense of Moscow reflected not only the constraints of the ABM Treaty, but also the traditional Kremlin expectation of self preservation against all reasonable odds.

Arms control and geopolitics: INF and future quandaries

Soviet lack of faith in the survivability of their own nuclear command-control systems was evident in their leaders' prompt and spirited reaction against NATO's deployment of INF (Intermediate Nuclear Forces), including Pershing II ballistic missiles and ground launched cruise missiles (GLCMs), in Europe beginning in 1983. NATO took this decision in response to Soviet deployments of intermediate range SS-20 land based missiles, beginning in the 1970s. Soviet objections to the P-II and cruise deployments were made on the basis that the SS-20s were not really a new kind of capability, but merely a modernization of previously deployed missile types. NATO judged that the Soviets were creating a new substrategic nuclear war fighting capability that would create a new rung on the ladder of escalation between theater and total war. SS-20s were mobile missiles with two MIRV warheads and an approximate maximum range of 4,400 km. The Soviet concern about NATO INF was mainly with respect to the Pershing II: its range (maximum about 1,800 km) and speed permitted it to strike at some targets in the western Soviet Union, including command bunkers, within ten minutes of launch.

Soviet leaders may have feared a NATO theater-strategic, counter-command first strike capability based in Europe and usable in response to their own intermediate systems, closing an open seam in the spectrum of deterrence and placing the onus on the Kremlin for escalation outside of Europe. If so, Soviet leaders were beginning to reason like their U.S. and allied NATO counterparts, comprehending the introverted logic of deterrence that could posit a substrategic nuclear war contained below the threshold of global war and despite NATO nuclear attacks directly into the Soviet Union itself. It seems more likely that Soviet leaders did not anticipate NATO's political reaction to the SS-20 deployments, nor the solidarity that held together in NATO governments despite Soviet intimidation and domestic political opposition to the "572" deployments (including boisterous nuclear freeze movements). But the Soviet fear of their command vulnerabilities to the P-IIs was real enough, and it begged the question how much faith they had in the consequences of an even more robust American nuclear attack. The Pershing-II missiles carried single warheads with an approximate maximum yield of 50kt (compared to the MIRVed SS-20s 250kt).

Geography, along with politics and military technology, had turned an incremental upgrade in military capabilities (from the Soviet standpoint) and an equivalent, tit for tat response to Soviet missile enhancements (from NATO's perspective), into a contretemps that was only resolved by the agreement signed by Presidents Ronald Reagan and Mikhail Gorbachev in December, 1987 to dismantle and destroy all of their "intermediate-range and shorter-range" ground launched ballistic and cruise missiles deployed worldwide (those with maximum ranges between 500 and 5,500 kilometers).[16] This agreement could be seen in retrospect as the most significant and daring

---

[16] This specification got around the difficulty that some technical experts referred to the Pershing II as a medium range ballistic missile and others as an intermediate range ballistic missile. For INF Treaty purposes, intermediate-range missiles were those with maximum ranges between 1,000 and 5,500 kilometers; shorter-range missiles, between 500 and 1,000 km.

accord reached by the Cold War nuclear superpowers. SALT and START negotiations understandably had a special cachet for arms control gurus and for media aficionados. But the INF Treaty was the first to require the elimination and destruction of an entire class of deployed nuclear weapons and launchers, including those located in the western Soviet Union and targeted on both Europe and Asia.

So pervasive and unprecedented was this Treaty that two decades later, Russian President Vladimir Putin reopened the discussion in Russia about whether to continue to adhere to the INF agreement. He noted in 2007 that only Russia and the United States had taken the drastic step of eliminating this entire class of missiles under very different political conditions than those that obtained in the early twenty-first century. Putin had a point. Both the U.S. and Russia worried about the spread of ballistic missiles, including those of medium and longer ranges, to states with revisionist goals, regional grudges and-or armed forces of dubious political reliability. Should the United States and Russia continue to practice INF chastity or self denial while others indulged their appetites for missile coercion at longer ranges – including missiles that might be tipped in the future with weapons of mass destruction, including possibly nuclear ones?

One of the things that had changed for Russia from 1987 to 2007 was its geography – compared to that of the former Soviet Union – especially by a western perimeter that was pushed eastward almost to Smolensk and Rostov-on-the-Don, as the non-Russian member states of the USSR gained their political independence. The sensitivity in Russia toward NATO or U.S. actions in their former Soviet space is not only political or military, but it is based on the hard facts of geography. Russia need not possess a "Barbarossa complex" about invasion from the West in order to notice that NATO's air-land conventional deep strike options are, on a map table or in PowerPoint deluxe, impressive relative to Russia's ability to trade space for time. The U.S. military campaigns against Iraq in 1991 and in 2003 were carefully studied by Russian military experts, tracking the "shock and awe" capabilities for dismantling any opposed conventional force playing by similar rules of engagement and operating within the same battle space. Indeed, the Americans' ability to turn almost any location on the planet into target coordinates in good time redefines battle space to the detriment of those who plan for a war of attrition, compared to a rapid battle of annihilation.

How much does the situation described in the previous paragraph actually matter for Russia? If the United States, NATO and Russia remain locked in a Cold War political world and in a similar military-strategic paradigm, then it matters a great deal, and the inferences for future Russian security are not very reassuring. However, policy makers are not automatons – the future is indeterminate – and politics is, or should be, the master faculty that rules over strategy. Therefore, the door is open for NATO and Russian political leaders to deconstruct old definitions of contested security space and reboot into another concept, of jointly shared responsibility for security in the North Atlantic and trans-Eurasian regions. Russia would be admitted into NATO (now a reborn Northern Treaty Organization, NTO) and further integrated into the economies of democratic Europe: ties that bind, to mutual advantage. Military retros in Russia who resist reform, on the grounds that NATO and the United States still constitute the major strategic threats to Russia, would be bypassed by the march of history. Issues as between Russia and the West would remain, but they would be resolved by diplomatic or other nonmilitary means, within a new security community from the Bering Sea to the Bering Sea.

The advantages to such a rethink of security, as related to geography, are several. First, the United States and Russia could bypass the otherwise distracting detour of coming clashes over the fault line territories between the former Russian and Soviet empires and the expanding geographical boundaries of NATO. Second, a U.S.–Russian and NATO–Russian security condominium would expedite cooperation across a number of nonmilitary issues related to security: including economics and natural resources; human rights; and the rule of law in democratic societies. Third, military to military discussions and consultations on issues of international security, as between NATO and Russia, could be expedited, including: nuclear arms control and nonproliferation; terrorism, including WMD-equipped malefactors, and insurgency; missile defenses and the military uses of space; and, the implications for NATO and Russia of rising regional military powers, including China and Iran. An additional benefit for modernizers in the Russian armed forces is that sideswipes from contact and cooperation with Western militaries would help to euthanize or convert forces resistant to reform. As Russia's armed forces became more comparable to, and interoperable with, the voluntary forces of (former) NATO members, momentum would increase in the direction of military professionalism in office and enlisted ranks, realistic general staff mission planning, and politico–military direction of the armed forces toward the cooperative Eurasian security tasks of the future (conflict prevention and resolution, peacekeeping, stability and security operations, and counterterrorism).[17]

---

[17] For alternative scenarios in U.S.–Russian future relations, see Aslund and Kuchins, *The Russia Balance Sheet*, pp. 146–148. Also informative are: Stephen J. Blank, *Russia and Arms Control: Are There Opportunities for the Obama Administration?* (Carlisle, Pa.: Strategic Studies Institute, U.S. Army War College, March 2009), and Dale R. Herspring, ed., *Putin's Russia: Past Imperfect, Future Uncertain* (Lanham, Md.: Rowman and Littlefield, 2007), Third Edition.

Against this, NATO and Russian skeptics will oppose two concerns: (1), NATO's fighting power will be dissipated into quasi-social enterprises and political great leaps forward, leading to mission malaise and lack of readiness for "real war" or necessary nuclear deterrent missions; or, (2), Russia's security will be compromised by absorption of its strategic uniqueness into a larger and riskier version of the OSCE. The heritage of Suvorov and Kutuzov will disappear into a sinkhole of postmodern demilitarization, rendering a future Russia vulnerable to new predations from the west, east, south and, perhaps, from space. And both NATO and Russian skeptics about Atlantic and trans-Eurasian political and military integration will ask hard questions about immediate political and military priorities: (1), what is the exit strategy or definition of victory for Afghanistan (and whose definition); (2), how drastically can the United States and Russia reduce their numbers of deployed nuclear weapons on intercontinental launchers, and can further reductions empower their future leadership to success on nonproliferation; (3), can the current situation with respect to U.S., NATO and Russian exploitation of space – its use for military purpose without deploying weapons in space – be continued in the face of possible challenges from other powers; (4), can the world economy support continued military modernization in the industrialized and postindustrial worlds without collapsing budgets for nonmilitary components of security (including the public management of health, education, natural resources and other ecological–environmental issues); and, (5), can competent strategies be developed for proactively anticipating and defusing the "culture wars" about ethnicity, religion and other primordial values that support war and terrorism within, and across, state borders?

These questions have no easy or obvious answers, but they support the case for adaptive, scenario-based planning that some have argued for as a higher profile activity for future Pentagon (and other) military leaders. U.S. Secretary of Defense Robert Gates has referred to the tendency within the American armed forces for preoccupation with future warfare to claim too much attention, compared to the demands of military operations of the present. Gates referred to this syndrome as "next war-itis." In response, some critics have warned that, as important as the war on terror or the wars in Afghanistan and Iraq may be, future conflicts may not repeat either the music or the libretto of current exigencies.[18] War is fundamentally a two sided, competitive affair: states and nonstate actors will design around the strengths and probe for the weaknesses of their opponents. Russia, for example, need not threaten Western Europe with military invasion or nuclear coercion:

---

[18] Andrew F. Krepinevich, 7 Deadly Scenarios: A Military Futurist Explores War in the 21st Century (New York: Bantam Books, 2009), p. 299.

Moscow can simply turn off the gas tap. In addition, not all catastrophes are purpose built – some result from the accumulation of low amplitude dysfunctions having reached a point of punctuated equilibrium. For example, Pakistan poses not only the threat of nuclear war between Pakistan and India, but also the threat of insurrection and regime change that leaves its nukes in politically dangerous, or unknown, hands.

Experts are often surprised in their anticipation of future war.[19] Therefore, prudent planners should not expect to avoid tactical, operational or even strategic military surprise. Instead, they should plan for minimizing the expected consequences of surprise and for responding decisively to the sources of menace. The relationship between nuclear weapons and geography (or nuclear arms control and geostrategy) also follows this logic. Some surprising things will undoubtedly happen with nuclear weapons in what remains of the present century, but those surprises will be built on precedents already known: the locations and ranges of nuclear delivery systems, their stationary or mobile locations, their electronic and personnel command-control systems, and, most important, their political ownership and military custodians. All of these important variables related to the likelihood and effects of future nuclear use, for deterrence or worse, are delimited by geographical and geostrategic factors. Even taking advanced non-nuclear weapons into space, in order to threaten nuclear weapons based on earth, involves the mastery of geospatial kinetic, electronic and cybernetic forces and powers.

At the other end of the spectrum, from space to the possibility of nuclear terrorism, nuclear strategy and geopolitics remain linked in an uneasy embrace. Terrorists, including possibly nuclear armed ones, are about the denial of geography: they cross borders and shift faux identities in order to strike at preferred targets. Nuclear weapons acquired by terrorists will, in all likelihood, have been fabricated by someone else. The terrorists' nukes will have been deliberately slipped to them by states or privateers, or terrorists will somehow have stolen them. However acquired, nuclear weapons in the hands of terrorists cannot be deterred from use. Those states that allow terrorists to steal nuclear weapons, or mercenaries within their borders to sell them, are as morally accountable for the consequences of terrorist WMD attacks as they would be if their governments had deliberately abetted nuclear terrorism.

This moral accountability should now be made into a legal one by the international community, with provisions for enforcement and consequences

---

[19] Colin S. Gray, *Another Bloody Century: Future Warfare* (London: Weidenfeld and Nicolson, 2005), pp. 29–54.

– including fines or jail terms for complicit officials. It will require the cooperation of the P-5 UN Security Council permanent members, as well as other nuclear weapons states, to accomplish this. But this new gold standard of accountability for fissile materials and weapons is a small price to pay, compared to the alternative. If governments cannot get this far, with respect to the risk management aspects of nuclear geostrategy, how much confidence can we have that they can master the larger challenges already discussed? Counterterrorism is as much a coordinate of credible nuclear geopolitics and geostrategy as is the heavenly gaze of star warriors, the fascination with advanced conventional blitzkrieg, or the toe-tapping resonance of network-centric warfare from the heavens above to the caverns of the Hindu Kush.

Conclusions – and further thoughts

One can imagine a number of alternative nuclear futures, including: (1), nuclear abolition; (2), nuclear marginalization, with greatly reduced arsenals compared to now; (3), nuclear abundance or plenty, with more nuclear weapons states and, probably, hair trigger weapons; (4), defense dominance, relative to offensive missile and-or airborne weapons; and, (5), complex uncertainty, revealing no single trend line but a potpourri of competing state, interstate and substate developments with respect to nuclear weapons and arms control.[20] This template of alternative nuclear futures is intersected by the seven dimensions of strategy, noted earlier. The result is a complicated mosaic for political leaders, military planners and scholars to grapple with here and now – never mind tomorrow.

The nuclear past already recorded provides some clue to the nuclear future and, therefore, to its relationship with geopolitics and geostrategy. Nuclear weapons did not disappear or even become marginalized with the end of the Cold War. They remain the most dangerous weapons of mass destruction, but political and bureaucratic inertia, combined with genuine security dilemmas, have moved history into a second nuclear age. In the second nuclear age, the world is no longer dominated by a political and military rivalry between two nuclear "superpowers": the United States and the Soviet Union. Through a combination of deterrence, military strength, containment, diplomacy and arms control, the first use of a nuclear weapon fired in anger since Hiroshima

---

[20] For an expansion, see Stephen J. Cimbala, *Nuclear Weapons and Cooperative Security in the 21st Century* (London: Routledge, 2010).

failed to occur during the first nuclear age.[21] Will a similar mixture of hard and soft power work to preserve a "nuclear taboo" in the twenty-first century, or will past tool kits be found wanting?[22]

Perhaps Martin Luther's injunction to "sin boldly" applies to social scientific and historically based efforts to guesstimate the future. However, the reciprocal relationship between nuclear technologies, on the one hand, and geopolitics and geostrategy, on the other, has a past, and from that some inferences follow. The geostrategic dimensions for future nuclear deterrence and arms control, or for war if deterrence and other prophylactics fail, are: land, sea, air, space and cyberspace. Of course, nuclear weapons cannot actually detonate in cyberspace, but deficient preparation for war or deterrence in cyberspace can create vulnerabilities exploited by a dedicated enemy: including an enemy armed with nuclear weapons. States operating nuclear forces will therefore have to plan for (at least) a five sided multidimensional behavior space, with respect to nuclear deterrence or antinuclear defenses.

We can also anticipate that antimissile defenses (or BMD, for ballistic missile defenses) and antiair defenses will improve, *relative* to the offensive weapons directed against them that date from the *first* nuclear age. The modern ballistic missile first appeared during World War II. Its ancestors date to antiquity: the Roman ballista or the rockets fired by Chinese combatants during the "warring states" period are exemplary. On the other hand, nuclear weapons in the twenty-first century will, unlike the situation in the first nuclear age, appeal to the militarily weak instead of the strong. Rogue states and nonstate actors will see nuclear and other weapons of mass destruction, together with missile and aircraft for delivery of WMD, as offsets or "asymmetrical" strategies as against the superior conventional militaries of the United States or others. A contradictory pattern may emerge, in which the major nuclear weapons states agree to freeze or cut their numbers of deployed warheads and launchers, while others see this as an opportunity to gain ground and

---

[21] Michael Krepon, *Better Safe than Sorry: The Ironies of Living with the Bomb* (Stanford, Calif.: Stanford University Press, 2009), passim. For other analyses of the second nuclear age (beginning at the end of the Cold War and the demise of the Soviet Union), see: Patrick M. Morgan, *Deterrence Now* (Cambridge: Cambridge University Press, 2003), pp. 238–284; Paul Bracken, *Fire in the East: The Rise of Asian Military Power and the Second Nuclear Age* (New York: Harper Collins, 1999), esp. pp. 95–124; Colin S. Gray, *The Second Nuclear Age* (Boulder, Colo.: Lynne Rienner, 1999); and Keith B. Payne, *Deterrence in the Second Nuclear Age* (Lexington, Ky.: University Press of Kentucky, 1996).

[22] On the concept of a nuclear taboo, see Nina Tannenwald, *The Nuclear Taboo: The United States and the Non-Use of Nuclear Weapons Since 1945* (Cambridge: Cambridge University Press, 2007), esp. pp. 327–360, and George H. Quester, *Nuclear First Strike: Consequences of a Broken Taboo* (Baltimore, Md.: Johns Hopkins University Press, 2006).

improve their relative nuclear status – including, in the latter group, regional rogues and their nuclear armed rivals.

Missile defenses may be tasked for deterrent and defense missions within a regional theater of operations more often than they are expected to deter, or defend against, attacks by transoceanic nuclear powers. Instead of the global nuclear war between superpowers that placed the entire planet in peril during the first nuclear age, we might see "small" nuclear wars or individual strikes confined within a single state or regional territory. The reaction of other nuclear weapons states to one or several isolated and discrete uses of tactical or operational nuclear weapons are not necessarily predictable from past practice or policy statements. Once the tradition of nuclear nonuse since Nagasaki had been broken and the fighting subsequently contained to serious, but less than catastrophic, levels for the affected societies, we would be in a new world of nuclear expectations. Indeed, it might not even be clear who had fired the first shot, or whether an accidental or unauthorized missile launch had triggered the episode in question.[23] If the nonproliferation regime proves to be nonresistant to regional nuclear proliferation, the appeal of counterproliferation strategies, combined with antimissile and antiair defenses, will increase.

Nuclear terrorism is the threat du jour, and U.S. or other national security policy makers and military planners cannot avoid taking it seriously.[24] However, terrorists of the future, as in the past, have obstacles to acquiring and using nuclear weapons. Terrorists cannot build a nuclear infrastructure or operate a nuclear fuel cycle without risking discovery and capture. Instead, they must rely on buying or stealing fissile materials and nuclear related technologies from other sources. These other sources can include states or entrepreneurs. The A.Q. Khan Pakistan-based network offered nuclear technology and know-how to a variety of interested state buyers – the jury is out with respect to terrorist customers. Terrorists, having acquired nuclear, chemical, biological or radiological materials, require safe havens for their plots. Those safe havens are located in states and they have specific geographical locales. With superior domestic security and just-in-time counterintelligence, these locations can be de-sanctuarized and, if necessary, destroyed. Terrorists, even armed with nuclear materials, can defy geography in a sense that states cannot – but they cannot escape or ignore the strategic vulnerabilities (or opportunities) created by geography. Outside of urban areas, there is a reason why terrorists prefer to hole up in mountains and jungles.

---

[23] For discussion of pertinent scenarios, see Quester, *Nuclear First Strike*, pp. 24–52.

[24] A timely and informative treatment of this subject appears in Graham Allison, *Nuclear Terrorism: The Ultimate Preventable Catastrophe* (New York: Times Books – Henry Holt and Co., 2004).

Can we envision a future without nuclear weapons or long range ballistic missiles and air forces to deliver them to deadly effect? Certainly we can, as some have done. The cliché that warns "Be careful what you wish for!" applies here. Nuclear weapons and their relationship to geography are two sided. If used, they have socially unacceptable and morally indefensible effects. If not, they can contribute to the avoidance of war by the credible promise of terror, mutually shared. However, this peace through terror pact only works if all parties understand it and share its rationality. As Patrick M. Morgan has noted, the Cold War and nuclear weapons "gave deterrence an undeserved good name" because in earlier times "the mixed utility of deterrence was much clearer in that it was often not very effective, sometimes counterproductive. That will be true now."[25]

The rationality of nuclear deterrence is not necessarily "western" – Mao understood it perfectly well. But it is statist. Nuclear deterrence is a West phalian construct par excellence. Only governments have the power to build and operate nuclear forces responsibly, for as long as nuclear weapons exist and are deployed. And only governments have the shared power to utterly destroy modern life through nuclear fire. Jonathan Schell suggests that nuclear catastrophe is analogous to global warming in the scale of the danger that it poses to civilization.[26] This comparison may privilege nuclear danger more than it deserves. Policy makers, existing in Rousseau's morally ambivalent civilization instead of his preferred state of nature, have to choose among options that are either more, or less, miserable. The debate with respect to alternative nuclear futures is reminiscent of Samuel Johnson's appraisal of the feud between Voltaire and Rousseau: "Why, sir, it is difficult to settle the proportion of iniquity between them."[27]

---

[25] Morgan, *Deterrence Now*, p. 263.

[26] Jonathan Schell, *The Seventh Decade: The New Shape of Nuclear Danger* (New York: Henry Holt, 2007), p. 7 and passim.

[27] James Boswell, *The Life of Samuel Johnson LLD* (London: John Murray, 1831), II, pp. 224–228, in Alan Ritter and Julia Conaway Bondanella, eds., *Rousseau's Political Writings* (New York: W.W. Norton, 1988), p. 202.

4

Nuclear abolition: Holy grail or dangerous temptation?

Introduction

Nuclear abolition is an idea whose time has come.[1] Prestige endorsements of the idea of nuclear abolition, or of drastic nuclear arms reductions, include U.S. President Barack Obama and Russian President Dmitry Medvedev. Addressing the United Nations General Assembly in September, 2009, Obama reiterated his earlier call in April, 2009 for a world without nuclear weapons, and Medvedev urged other nuclear states to "join the disarmament efforts of Russia and the United States" without waiting for "further progress in the

---

[1] For example, see George P. Shultz, William J. Perry, Henry A. Kissinger and Sam Nunn, "Toward a Nuclear-Free World," *Wall Street Journal*, January 15, 2008, p. A13. See also: Henry A. Kissinger, "Containing the fire of the gods," *International Herald Tribune*, February 6, 2009, http://www.iht.com/articles/2009/02/06/opinion/edkissinger.php; Press Association, "David Miliband sets out six-point plan to rid world of nuclear weapons," *guardian.co.uk*, February 4, 2009, http://www.guardian.co.uk/politics/2009/feb/04/miliband-nuclear-weapons; and Jennifer Loven, "Obama outlines sweeping goal of nuclear-free world," *Associated Press*, April 5, 2009, in *Johnson's Russia List* 2009 – #66, April 5–6 2009, davidjohnson@starpower.net. For an especially insightful commentary on this topic, see Thomas C. Schelling, "A world without nuclear weapons?" Daedalus (Fall, 2009), No. 4, pp. 124–129.

Russian–American disarmament process."[2] As in the case of past efforts to abolish or marginalize the role of nuclear weapons in national strategies and foreign policies, the current interest in abolition is motivated by the best of intentions.[3] However, the management of global decision-making processes toward the elimination of nuclear weapons is a challenge that has escaped resolution since the first detonations in 1945. This chapter reviews some of the difficulties in the way of realizing nuclear abolition in fact, as opposed to a demonstration of its presumed desirability in theory. We also consider whether important arms control objectives, short of nuclear abolition, may also be attainable, or even imminently feasible.

The high concept of nuclear abolition admits of many, perhaps too many, operational definitions

By nuclear "abolition" one might intend to convey a proposal in favor of any one of the following: elimination of all deployed nuclear weapons; elimination of all deployed and stored nuclear weapons; the preceding, plus international control of weapons grade materials and-or nuclear fuel cycles; all of the above, plus dismantling of manufacturing and construction facilities, and perhaps laboratories, deemed suspect or renegade by appropriate international authority; and, all of the preceding, plus the availability of armed forces "on call" to international authority for the purpose of preemptive or preventive intervention to disarm rogue states or nuclear terrorists.

The Obama administration's *National Security Strategy* published in 2010 vows to pursue the goal of a world without nuclear weapons – although, so long as nuclear weapons still exist, the United States will maintain an effective, safe and secure nuclear arsenal to deter possible opponents and

---

[2] President of Russia Dmitry Medvedev, Address to the 64th Session of the UN General Assembly, New York, September 23, 2009, *Kremlin.ru*, in *Johnson's Russia List* 2009 – #177, September 24, 2009, davidjohnson@starpower.net.

[3] For example, see "At UN, deadline aired for abolishing nuke weapons," *Associated Press*, http://www.google.com/hostednews/ap/article/, downloaded May 19, 2010.
For contrasting perspectives on this issue, see Ivo Daalder and Jan Lodal, "The Logic of Zero," *Foreign Affairs*, No. 6 (November–December, 2008), pp. 80–95; and Colin Gray, "To Confuse Ourselves: Nuclear Fallacies," Ch. 1, pp. 4–30; Lawrence Freedman, "Eliminators, Marginalists, and the Politics of Disarmament," Ch. 4, pp. 56–69; and Michael MccGwire, "The Elimination of Nuclear Weapons," Ch. 9, pp. 144–166, all in John Baylis and Robert O'Neill, eds., *Alternative Nuclear Futures: The Role of Nuclear Weapons in the Post-Cold War World* (Oxford: Oxford University Press, 2000).

to reassure allies and other security partners.[4] An ambitious U.S. agenda of steps intended to reduce nuclear danger includes the following:

> We are reducing the role of nuclear weapons in our national security approach, extending a negative security assurance not to use or threaten to use nuclear weapons against those nonnuclear nations that are in compliance with the NPT [Nuclear Non-Proliferation Treaty] and their nuclear nonproliferation obligations, and investing in the modernization of a safe, secure, and effective stockpile without the production of new nuclear weapons. We will pursue ratification of the Comprehensive Test Ban Treaty. And we will seek a new treaty that verifiably ends the production of fissile materials intended for use in nuclear weapons.[5]

Many persons speak of nuclear "abolition" when they actually mean to refer to nuclear disarmament. They recognize that the world cannot return to a time before nuclear knowledge existed, and they doubt that states would voluntary give up sovereignty to the extent necessary for the high end of the "abolition" curve. Nuclear disarmament sounds more practical, although daunting in its own right, than nuclear abolition. To "disarm" implies the removal or destruction of actual weapons from the arsenals of states. Disarmament has a track record in history. Both conventional and nuclear disarmament have been accomplished by international agreement, as in arms control negotiations, by self abnegation, when a state voluntary gives up arms, or by imposition, when outsiders coerce or compel surrender of weapons. The last option frequently follows on the heels of military defeat.

Nuclear arms control grew up as a preferred means of limitation during the Cold War because disarmament failed to appeal to the United States, to the Soviet Union, and to other members of the original nuclear club (also the five permanent members of the UN Security Council). Arms control seemed less threatening to besieged policy makers and military planners with defense problems to solve than did the idea of disarmament. In addition, U.S. approaches to Cold War arms control emphasized that the negotiation of arms limitations had to take into account more than the numbers of nuclear weapons. The qualities of weapons deployed or withheld, and their psychological impacts for nuclear stability, were judged to be of primary significance for the avoidance of war. Influential early Cold War American and other strategic thinkers emphasized that the major purpose of nuclear weapons after Nagasaki

---

[4] The White House, *National Security Strategy* (Washington, D.C.: The White House, May, 2010), p. 23.

[5] Ibid.

would be the avoidance of war, not the achievement of military victory in the traditional sense. Therefore, the appeal of "deterrence" as an experiment in applied psychology, relative to the maintenance of nuclear stability and the avoidance of nuclear war, became seminal. By the early 1960s both U.S. and Soviet leaders were faced with the problem of managing Wohlstetter's "delicate balance of terror" by the manipulation of the risk of war in order to avoid war. Schelling's "threat that leaves something to chance" recognized the reciprocal responsibility of both sides for a failure of deterrence, in that they shared the power to pull both contestants over the rope and into the mud beyond it.[6]

With hindsight, deterrence "worked" well enough during the Cold War, at least from the American and allied NATO standpoint.[7] We had the right number of nuclear wars: none. We had the right number of major wars in Europe: none. And the United States and NATO prevailed politically at the endgame of the Cold War, when the Soviet Union collapsed without kicking off a Third World War with its death rattle (a consequence by no means impossible, given the indeterminacy of the outcomes of the upheavals in Moscow from August 19 to 21, 1991). However, and paradoxically, these "successes" were not necessarily or irrefutably caused by finesse in the practice of deterrence.

Instead, it is equally likely that the avoidance of World War III during the first nuclear age was overdetermined: by the remembered devastation of World War II that no one wished to repeat, with or without nuclear weapons; by the lack of any dispute between the Soviets or the Americans over contiguous heartland territory; by the ambiguity whether, and to what extent, NATO would hang together in the face of Soviet intimidation or war (always scenario-dependent); and, as counterpart, whether the outbreak of major war in Europe would provide the opportunity for revolt against Soviet domination by the states of Eastern and Central Europe included in the Warsaw Treaty organization. The political contexts within which any nuclear war would have to be fought, as opposed to deterred or otherwise avoided, pushed national

---

[6] Thomas C. Schelling, *Arms and Influence* (New Haven: Yale University Press, 1966), p. 121, note 8.

[7] Assessments of deterrence before and after the Cold War appear in: Patrick M. Morgan, *Deterrence Now* (Cambridge: Cambridge University Press, 2003); Colin S. Gray, *The Second Nuclear Age* (Boulder, Colo.: Lynne Rienner, 1999); Keith B. Payne, *Deterrence in the Second Nuclear Age* (Lexington, Ky.: University Press of Kentucky, 1996); Robert Jervis, *The Meaning of the Nuclear Revolution: Statecraft and the Prospect of Armageddon* (Ithaca, N.Y.: Cornell University Press, 1989); and Lawrence Freedman, *The Evolution of Nuclear Strategy* (New York: St. Martin's Press, 1981 and 1983. Michael Krepon emphasizes that deterrence in the first nuclear age "worked," to the extent that it did so, only in conjunction with containment, diplomacy, military strength and arms control. See Krepon, *Better Safe than Sorry: The Ironies of Living with the Bomb* (Stanford, Calif.: Stanford University Press, 2009), passim.

leaders toward nuclear restraint and against excessive faith in nuclear deterrence and brinkmanship. As Michael Krepon has noted:

> There was no greater safeguard during the first nuclear age than the distance between the plans of nuclear strategists and the political instincts of national leaders. The separate worlds of the weapon strategists and political leaders became the most important fail-safe mechanism to prevent a nuclear holocaust.[8]

Overdetermined or not, relative peace in Europe from 1946 to the end of the century, compared to the first forty-five years of the same century, had something to do with the existence of nuclear weapons; and not only the existence of weapons – but also, the numbers and varieties of nuclear capable launch systems, the preparation of detailed war plans, and the rehearsal of command and control procedures in the event that deterrence failed. These exercises and plans, at least on the U.S. side, sometimes took on a surreal aspect, with aspirations for post-Armageddon skirmishing by surviving remnants of depleted Soviet and American nuclear arsenals, commanded by ghostly apparitions from bunkered bivouacs. Nevertheless, the Americans and the Soviets took seriously the possibility of an outbreak of nuclear war and prepared for the worst. Arguably, these preparations, grisly as they were, strengthened the credibility of deterrence that made possible the bypass of nuclear destruction between the end of World War II and the end of the Cold War. Deterrence was a substitute for abolition, although not an adequate one in the minds of those who favored more complete nuclear disarmament.

Politics rules – for better or worse!

This resumé of past experience has a point for future advocates of nuclear abolition, however they define the term. Nuclear weapons in the Cold War acted more as pacifiers than they did as contributors to the onset of war and mass destruction. They did so because politics rules over military strategy, including arms control.[9] The two nuclear armed "superpowers" of the Cold War found

---

[8] Krepon, *Better Safe than Sorry* p. 55.

[9] The point is made with appropriate emphasis in Gray, *The Second Nuclear Age*, pp. 17–46, who notes that: "From a Cold War condition of notable, if understandable, underachievement, the scholarly authors of large organizing ideas have been working overtime since the USSR closed for business at the end of 1991. The problem is that from a condition in the 1980s wherein there was too little competitive theorizing of a basic kind about alternative political contexts, today we probably have too much of such theory." (Ibid., p. 28).

that they had to contest for global influence beneath the nuclear threshold. As U.S. President Ronald Reagan and Soviet leader Mikhail Gorbachev eventually made official, a nuclear war cannot be won and should never be fought. On the other hand, the capacity for making credible threats of nuclear first use or retaliation outlived the Cold War and created a second nuclear age. This second, post-Cold War nuclear era offered an international system that was potentially less well fitted for compatibility with nuclear weapons than the Cold War was. When the first President Bush called for a "new world order" in the aftermath of the end of the Cold War, he envisioned a solidarity among existing nuclear weapons states that would prove to be elusive. In addition, he could not have foreseen the threats that would appear over the horizon, including early in the next century.

International politics, by the time of the second President Bush, had moved well beyond the constraints of Cold War political alignments. Relations between the United States and Russia were officially nonhostile, although tinged with feelings of competitive envy on the part of Putin's Russia. Instead of the focused threat of the Cold War, Washington and NATO were now faced with diffuse threats from various quarters. Among these dispersed and uncertain sources of threat were the nonstate actors, including transnational terrorists like al-Qaeda, capable of inflicting strategic harm on the United States and its allies. Two aspects of these post-Cold War terrorist organizations were chilling for U.S. and allied defense planners. First, jihadist terrorists were sometimes motivated not only by concrete objectives, but also by apocalyptic or chiliastic visions that included their own martyrdom. Their demands were as non-negotiable as their tactics were brutal and their hatred, feral.

A second aspect of the "new terrorism" of the twenty-first century was the possibility that terrorists or other irregular warriors might acquire weapons of mass destruction (WMD) – including chemical, biological, radiological, or nuclear weapons. Although none of these possibilities was encouraging, the acquisition or use by terrorists of nuclear weapons was the most discouraging possibility. The civilized world reacted in horror to the attacks of 9/11, not only on account of the destruction actually accomplished, but also because of the residual fear of "what next?" And "next" might very well be an exploding nuke in a major American or allied city. Nuclear learning was so widely distributed by the dawn of the millennium that terrorists could acquire a nuclear weapon by getting hold of fissile material and assembling it themselves, or by stealing or otherwise acquiring an assembled weapon from a nuclear weapons state or a black market supplier.

These fears of nuclear terrorism were realistic, and all the more threatening because nuclear terrorism reboots the politics of nuclear war and deterrence. Terrorists have no central government or "state" to defend. Unlike states, terrorists are organized along the lines of networks instead of hierarchies. Therefore, they present a poor target for nuclear deterrence or coercive diplomacy as practiced during the Cold War. In addition, their motives, if based upon visions of the end-of-days instead of the end of regimes or governments, elide the cost-benefit calculus of most deterrence models. Recognition of these and other possible limitations in applying deterrence to nonstate actors, the George W. Bush administration turned to the option of preemption or preventive war to disarm regimes that might harbor terrorists or, when possible, to strike directly at the terrorists themselves.

Preemption and preventive war are alternate forms of anticipatory attack.[10] Preemption assumes that an opponent has already set in motion an attack, or is about to do so. His decision is irrevocable. Preemption is therefore a decision to strike first in order to limit the damage that would result, compared to striking second. Preventive war, on the other hand, anticipates a future attack by another state or entity but is not based on intelligence that an enemy attack is actually under way, or has actually been irrevocably decided upon. By way of example, Israel's attack on Egypt in 1967 was a case of preemption. In the early years of the Cold War, various arguments were made by U.S. leaders for preventive war against the Soviet Union before the USSR became a peer competitor in nuclear capability. The distinction between preemption and preventive war can be indistinct at the margin and submerged in the documentation: depending on the historical case. The July crisis of 1914 provides illustrations of reasoning in favor of preventive war or preemption by political leaders and military planners among the Entente and Alliance powers.

The Bush II administration found it difficult to act preemptively against terrorist networks, although one could hunt them down and destroy them after the fact. However, preemption against states that might be suspect of sponsoring or harboring terrorists was another option. The most visible and controversial case in point was, of course, Iraq under Saddam Hussein. United States and allied intelligence estimated in 2002 and 2003 that Iraq had substantial quantities of WMD still undiscovered by UN inspectors. The possibility that Iraq might have a reconstituted nuclear development program, despite its apparent destruction after the Gulf War of 1991, could not be ruled out by intelligence estimators. Based on these now-controversial assessments, the Bush II administration opted for war and regime change in Iraq. Bookshelves groan as to whether this was a war of choice or a war of

---

[10] Anticipatory attack is discussed in Karl P. Mueller, et. al., *Striking First: Preemptive and Preventive Attack in U.S. National Security Policy* (Santa Monica, Calif.: RAND, 2006), and Colin S. Gray, *The Implications of Preemptive and Preventive War Doctrines: A Reconsideration* (Carlisle, Pa.: Strategic Studies Institute, U.S. Army War College, July 2007).

necessity, but that judgment is not the issue here. For present purposes, it suffices to note that, faced with a possible threat of WMD use by Hussein or allied terrorists, Bush opted for preventive war (mistakenly called preemption by some) against Iraq.

The George W. Bush decision to invade Iraq and overthrow Saddam Hussein became controversial, not to say white hot, in American politics after postwar weapons inspectors found no evidence of any WMD in Iraq. Critics charged that Bush and his advisors deliberately politicized and biased prewar intelligence estimates to inflate the threat of WMD and any Saddam–terrorist connections. There is within this episode, to be sure, a warning about the dangers of confusing intelligence estimation with marketing and sales. Nevertheless, the U.S. decision for preventive war was not made with the hindsight of postwar inspections: policy makers were obliged to depend upon fuzzy sets and imperfect indicators of the "perhaps" variety. Anyone who has dealt with intelligence estimations in the American or other governments with large intelligence communities will report that national "estimates" are almost always based on a mixture of fact and conjecture. And the quantity of conjecture is higher as the stakes inherent in the decision rise: for those making it, and for others who will feel the consequences.

Seeing history in retro through the lenses of Bush II decision makers provides a case study in support of the heading for this section. As political actors, intentions and policies change, strategy must follow. Bush II strategy may have succeeded or failed in Iraq, or with regard to the war against al-Qaeda still in progress. However, the conjunction of rogue states and outlaw movements with the possibility of shared WMD is a realistic, and suitably frightening, one for United States and other twenty-first century military planners. Worse is the recognition by counterterror and counterinsurgency experts that both of these forms of conflict begin "at home." The internal security of states (homeland security in the U.S. vernacular) is the front line in the war against transnational or other terrorist groups. Thus the United States and NATO cannot impose success in counterterror or in counterinsurgency on other governments (unless regime change is on the menu, but even then, difficulties arise – viz. Iraq from 2003 to 2006), but must work with sometimes recalcitrant regimes to improve their domestic intelligence and police work.

The case of Iraq from "Mission Accomplished" to the fall of 2006 also shows that the military aspects of strategy, important as they are, are insufficient by themselves to resolve the turbulence of civil war and humanitarian disasters in failed states or otherwise troubled polities. As U.S. General David Petraeus said, with reference to U.S. military operations in Iraq and the objective of a stable and democratic regime, "We cannot kill our way out of this problem." The political, cultural, socio–economic and other "soft" aspects of Iraqi society and U.S. influence would weigh heavily in the balance. So, too, would the "battle of ideas," between advocates of democracy and proponents of authoritarianism that might contribute to terrorism. The competition in soft power, including ideas, requires of U.S. forces a combination of military, diplomatic, psycho–social, cultural and other skill sets that their grandfathers in arms might never have acknowledged, let alone practiced. If Shakespeare's admonition that "we must be cruel only to be kind" seemed to apply to mass industrial warfare with its large amounts of collateral damage, the reverse epigram might be better used for the "new war" or "long war" against terror networks insurgents – the counterterror side must be kind (to the general population, by providing security and the means for reconstruction of economies and societies) in order to be cruel (to the enemy).

What has the preceding discussion have to do with nuclear danger? Two things: one fairly obvious, and one relatively less so. The more obvious point is that terrorists armed with nukes require increased seriousness on the part of states in favor of nuclear containment and nonproliferation – whether those states are committed to nuclear abolition or simply to the avoidance of nuclear terrorism. Second, and relatively less obvious, is that the creation of additional nuclear weapons states, although not necessarily leading to the diffusion of nuclear technology and know-how into the hands of terrorists, certainly increases the likelihood of such an outcome. The second point invites misunderstanding, addressed in the following section.

Weapons do not cause wars – although they are, on occasion, enablers of mistaken strategies

Weapons are dependent variables, more often than not. States go to war, as Thucydides noted, for motives of fear, honor and interest. States arm because they fear for their security in an international system of endemic insecurity. This systemic condition cannot be erased by world government, by benign hegemony on the part of any single great power, or by prudence born of fear about nuclear self destruction – despite the hopes pinned on each of these "remedies" by their proponents. At best, a balance of power among rival states and across different issue areas (political, military, economic, social, and so forth) maintains whatever systemic stability and predictability exists. These balances of power fluctuate though time and geostrategic space with respect to their ability to induce or to compel system-stabilizing behavior. Empires are not exceptions to the preceding generalizations. Some empires adapt to changed conditions better than others, but all social systems are inherently entropic, and the international system offers the temptation of self destruction by means of imperial overbite.

The idea that nuclear weapons can be the causes of war is not mistaken in its entirety. It has the status of a valid warning about the decision-making processes within nuclear armed states. Once having acquired a nuclear arsenal, states have every reason to want to maintain authoritative legal, and effectively secure, command and control of their nuclear forces. Politico–military command and control over nuclear weapons must accomplish at least three things: (1), nuclear weapons must not be launched or detonated accidentally; (2), nuclear forces must be timely responsive to duly authorized commands; and, (3), peacetime nuclear custody must be proof against corruption of security, including theft or loss of nuclear materials or weapons.

Every state that has chosen to develop and deploy nuclear weapons has had to deal with the checklist of missions noted in the preceding paragraph. The Cold War Americans and Soviets, over several decades of trial and error, developed processes and procedures to ensure that command and control functions were performed according to the requirements of policy and of military effectiveness. Given the differences between their respective political systems, the United States and the Soviet Union prepared to manage their forces somewhat differently in time of peace, crisis and war. Nevertheless, the uniquely destructive character of nuclear weapons and the prodigious size of their respective arsenals mandated that, at least in dealing with one another, prudence was mandated in the management of nuclear operations. The unfolding of the Cuban missile crisis was both a demonstration of this point, and a syllabus of instruction for U.S. and Soviet leaders about the potential for nuclear crisis management to escape their shared control.

Today, most foreign offices worry less about another Cuban missile crisis than they do about the possibility of "loose nukes" in states with large numbers of stored or deployed weapons or significant quantities of weapons-grade material. These weapons or materials could, as previously noted, find their way into the toolkits of terrorists or rogue states, including those inured to influence by deterrent threats. Even the post-Soviet Russian nuclear arsenal and weapons complex is a source of concern in this regard. Under the Nunn–Lugar program, the United States has provided assistance to Russia for securing or dismantling nuclear weapons and materials for almost two decades. How loose the nukes in nuclear weapons states with fewer years of experience in storing, guarding, transporting and managing nuclear weapons might be, is a matter for considerable argument. Pakistan, for example, has apparently taken promising steps to make its weapons more secure and has gone some way to provide transparency in this regard, for the benefit of the UNITED STATES and other concerned parties. On the other hand, skeptics fret that the problem in Pakistan (and elsewhere) is not so much the lack of technical safeguards as it is uncertainty about the political reliability, or lack thereof, of key players in the policy process – including the military and intelligence community.

In the case of North Korea, a situation of considerable opacity exists with respect to the actual nuclear command and control procedures that are followed under normal peacetime conditions – let alone during a crisis or war. The predominant fear of outside observers is not the inadvertent loss or theft of nuclear materials or weapons. Instead, concern rests on the potential for North Korea to become a supplier state for other state and nonstate actors who want to acquire nuclear capability or know-how. These concerns are well grounded. The Israelis bombed a construction site in Syria in 2007 that was assessed by U.S. intelligence as a reactor complex being built with support from North Korea. Once completed, the complex might have provided nuclear weapons-grade materials to Iran or other interested buyers. North Korea is also a leading exporter of ballistic missiles and missile technology, and past beneficiaries have included Pakistan, Iran and Libya.

North Korea's own nuclear weapons and ballistic missile programs are additional reasons for UNITED STATES and other suspicions about its intentions. The DPRK's nuclear capabilities have been the subject of six-sided diplomatic parleys including North and South Korea, the United States, Russia, China and Japan. The objective of the George W. Bush and Barack Obama administrations was to obtain the complete, verifiable and irreversible destruction and dismantlement of North Korea's nuclear weapons complex. However, it was obvious that this would not be accomplished, if at all, in a fortnight. Negotiations with North Korea are, even under the best of conditions, protracted and arduous. The benefits of a successful outcome are significant, and the risks of failure, not inconsiderable. If North Korea becomes an irreversible nuclear weapons state, the possibility that Japan and South Korea will also go nuclear cannot be excluded. The nonproliferation regime might fall over backward from Asia, rippling into the Middle East or even into Europe. Or so, at least, is the stated fear of nonproliferation experts.

On the other hand, if we assume that the North Korean political leadership is thinking strategically, then Pyongyang's best option might be nuclear opacity and indeterminacy. North Korea would dismantle part, but not all, of its nuclear manufacturing complex under international agreement, while retaining some manufacturing and stored materials that were "off limits" to international inspectors. This modus vivendi would trouble those who distrust the DPRK leadership based on its past behavior, but it has a quasi-precedent in the 2008 agreement between the UNITED STATES and India. Admittedly, India is a democracy, is more transparent than North Korea about its international behavior, and is now an American "strategic partner" in Asia in view of its potential for balancing against China. But a partial solution for North Korea might be the best that the Gang of Six can obtain, and better than the continuation of the present impasse.

The case of Iran makes the point, as does the situation in North Korea, that the character of regimes and of domestic policy-making processes cannot be separated from the issue of nuclear danger. Additional nuclear weapons states present not only a risk of deterrence failure and war but also a greater potential for nuclear weapons spread.[11] We have seen above that the "weapons don't make war" argument can be overstated: insecure weapons or materials can leak to the bad guys, and unreliable militaries armed with nuclear weapons could possibly commit an act of unauthorized launch with deadly effect. But these arguments reinforce the point that the "who" of nuclear ownership is as important as the "what" of nuclear weapons. Henry Sokolski makes this point, perhaps with undue pessimism, as follows:

> Before 2020, the United Kingdom will find its nuclear forces eclipsed not only by those of Pakistan, but of Israel and of India. Soon thereafter, France will share the same fate. China, which has already enough separated plutonium and highly enriched uranium to triple its current stockpile of roughly 300 nuclear warheads, could expand its nuclear arsenal too. Meanwhile, Japan will have ready access to thousands of bombs worth of separated plutonium . . . Compounding these developments, even more nuclear weapons-ready states are likely: as of 2009, at least 25 states have announced their desire to build large reactors – historically, bomb starter kits – before 2030.[12]

The cynic's retort to the preceding discussion, about the significance of the nuclear "who" for international nuclear deterrence and crisis stability would be that capabilities are what count: intentions can change. The United States and other states must arm to do their worst against the most capable possible challenge that they might have to face. On the other hand, the "capabilities are what count" argument could be restated with mischief as "capabilities are what can be counted." Intelligence organizations during the Cold War were fixated on numbers of missiles and launch systems to the sometime detriment of more reliable knowledge about intentions. This proved to be the case in the

---

[11] Expert argument to this point appears in Henry Sokolski, "Moving Toward Zero and Armaged-don?," Ch. 5 in Sokolski, ed., *Reviewing the Nuclear Nonproliferation Treaty (NPT)* (Carlisle, Pa.: Strategic Studies Institute, U.S. Army War College, May 2010), pp. 77–101.

[12] Ibid., p. 77.

Cuban missile crisis of 1962. Among other surprises, U.S. knowledge of its apparent nuclear superiority relative to the Soviet Union, deliberately leaked to the press in 1961, had the effect not of deterring Moscow from adventurism, but of the opposite. Khrushchev's bold gamble in Cuba was partly motivated by a desire to reframe the perception of Soviet nuclear inferiority by a knight's move of Soviet nuclear forces into Cuba. As in the case of the American fleet at Pearl Harbor in 1941, one man's deterrent was another's instigation to design around it.

Does deterrence still matter, and if so, how?

The George W. Bush administration, perhaps overly impressed by the apparent novelty of the threat of unconventional warfare using weapons of mass destruction, ushered in a premature funeral service for deterrence. Deterrence had grown up in the context of the Cold War and nuclear weapons, especially focused on the bipolar competition between the United States and the Soviet Union. However, the ambit of deterrence as a concept was broader than the Cold War or the U.S.–Soviet rivalry. Deterrence expressed the understandable desire of states in the first nuclear age to attempt to achieve their policy objectives without nuclear war, but also without conceding any vital interest in the process of coercive bargaining. Nuclear deterrence provided a compromise between deterrence and Armageddon, albeit a Faustian bargain to some of its critics.

Critics who regarded nuclear deterrence as unacceptable were not confined to the political left. Many military traditionalists preferred deterrence by denial, based on the capability of conventional forces to deny the opponent his objectives. The problem with deterrence by denial based exclusively on conventional forces was that, even before the nuclear age, it had a mixed track record. States repeatedly opted for war against enemies with superior capabilities, including those with larger and better trained forces, superior technology and (worst of all) a way of war that was one generation ahead of its contemporaries. Since the Melians refused to surrender to the Athenians, polities have defied odds or persevered in the face of them for reasons of state, for glory, for booty or simply on account of cultural predispositions to fight. The historical record in favor of conventional deterrence prior to 1945, as well as afterward, was not one to inspire optimism or overconfidence. As Ariel E. Levite has noted, Israel's concerns about the possibility of giving up its nuclear option in a nuclear-disarmed world include the uncertain credibility of conventional deterrence in such a context.

It may bring the world back to the business of calculating conventional military balances, a notoriously unreliable and risky business. It could unleash new security anxieties that in turn might generate a highly expensive conventional arms race, one that Israel could ill afford if it wishes to remain a prosperous and appealing society.[13]

Nevertheless, some expert and lay opinion was still so horrified by the collateral damage attendant to even small nuclear wars that preferential treatment of conventional deterrence was argued for, if at all possible. NATO was faced with difficult decisions about the relationship between conventional and nuclear deterrence throughout its Cold War history. Europeans wanted the American nuclear umbrella not only as a deterrent against a Soviet nuclear attack on Western Europe, but also as a guaranty against any outbreak of major war in Europe. From their side of the Atlantic, many NATO European defense ministers and foreign offices regarded U.S. nuclear weapons deployed in Europe as coupling devices that would preclude separation of "war in Europe" from global war. The assumption was that the Soviet Union, faced with a stark choice between "no war" and total war, would opt for no war even if it meant the surrender of vital interests.

Beginning with the Kennedy–Johnson administration and the American movement toward a flexible response strategy, the United States preferred to mix conventional with nuclear deterrence and, in the process, to establish in the minds of some Europeans the fear of a "firebreak" between war in Europe and total war. This prospect Europeans regarded with apprehension, allowing the Soviet Union to divide the alliance and bite off parts of it without risking total war and societal destruction. The U.S. Under Secretary of Defense Robert S. McNamara sought to improve both conventional defense and nuclear retaliatory forces, including those deployed in Europe. The U.S. assumption was that a graduated deterrent instead of an all or nothing choice between defeat and doomsday was more believable to Moscow, and therefore more successfully deterring.

At the center of the trans-Atlantic fibrillations about whether graduated or eruptive deterrence was more credible in Moscow was the other side of the coin: how believable was it in Brussels, or in Washington? NATO never deployed sufficient numbers of conventional forces in Europe for a protracted war of the duration of World War II. Europeans refused to pay the defense budgets that such a force would have required, and the UNITED STATES had neither the personnel nor the financial means to provide comprehensive

---

[13] Ariel E. Levite, "Global Zero: An Israeli Vision of Realistic Idealism," *The Washington Quarterly*, April 2010, pp. 157–168, citation p. 166, http://www.twq.com/10april/docs/10apr_Levite.pdf.

conventional defenses for Europe. In addition to the apparent hopelessness of preparing to fight World War III with conventional weapons, the nuclear genie could not be put back into the proverbial bottle after 1945. Once the Americans and Soviets had deployed large, diverse and presumably first strike survivable long range nuclear forces, it seemed superfluous to prepare for a replay of the Battle of Kursk on the central front in Europe.

NATO conventional forces, including American ones, forward deployed in Europe during the Cold War therefore served a purpose other than the traditional definition of conventional deterrence by denial. In fact, they served several very important political purposes that otherwise contributed to conventional and to nuclear deterrence.[14] First, forward deployed American "GIs" would be killed in the earliest stages of any Soviet attack across the Fulda Gap or other probable access corridors from East to West. These casualties would become martyred heroes in the minds of an outraged American public, thus susceptible to appeals for military escalation – including, if necessary, the use of nuclear weapons in retaliation. Second, UNITED STATES and allied European conventional forces provided a flexible capability for managing and containing a diplomatic crisis short of war – as in Berlin during the late 1940s, late 1950s and early 1960s. Conventional forces could, under some circumstances, give more credible support to coercive diplomacy than nuclear forces could – because conventional forces could be positioned into delicate spots, requiring the Soviets to accept the "last clear chance" to avoid a war.

For example, Khrushchev regarded the allied presence in West Berlin as a bone in the throat of the USSR, but also as a vulnerable point for diplomatic demarches and political bluster, supported by the shadow of Soviet military power nearby. His rhetoric in this regard was often menacing, but his military steps were measured. While some of the ebullient Soviet leader's cautious behavior can be explained by the existence of American and NATO nuclear capabilities, Khrushchev also had to reckon with the risks created by NATO's more-than-ceremonial ground, maritime and tactical air forces deployed in Western Europe. This was especially true in a divided Germany with a truncated Berlin nested within its eastern, and Soviet-controlled, sector. A seizure of Berlin by Soviet forces would eventually overpower American and allied conventional forces there, but not before the escalation of fighting had led to significant casualties, televised civilian pandemonium on both sides of the Iron Curtain, and hasty discussions among military planners and their political principals about nuclear release and possible first use.

---

[14] Bernard Brodie's discussion of the relationship between conventional denial and nuclear deterrence in NATO Europe has not been improved upon. See Brodie, *War and Politics* (New York: Macmillan, 1973), pp. 375–432, esp. pp. 396–412.

In other words, as George H.W. Bush once said, quoting Woody Allen, "Ninety per cent of life is showing up." NATO conventional forces in Europe, including those in Germany, were a more than public relations obstacle to Soviet military adventurism – even if a crisis were contained short of nuclear first use. The evident possibility of nuclear first use by either side changed the symbolism of the first use of conventional weapons fired in anger. Under the shadow of possible but uncertain nuclear escalation, each hour and minute of conventional war fighting, however localized, became a burning fuse leading to an unmanageable explosion. This uncertainty, about the degree and pace of escalation between two sides caught up in a process over which they might eventually lose control, was the centerpiece of Cold War deterrence. Nuclear escalation was neither definitely precluded nor assuredly guaranteed; instead, it was circumstantially dependent, and the circumstances were nonlinear and probabilistically pathological. Rumsfeld's "unknown unknowns" abounded.

In short, neither nuclear nor conventional deterrence during the Cold War was as simple or one dimensional as has been portrayed in many accounts. Nuclear deterrence then and now was, and is, multidimensional, nonlinear and subjective in many of its aspects. It was not merely, or mainly, about toting up numbers of weapons and launchers and punching them through computer programs in order to determine hypothetical "winners" and "losers" in a nuclear war. Instead, Cold War deterrence, including the nuclear aspect of it, relied on designed ambiguity, deliberately created uncertainty, and bounded rationality with respect to the challenge of understanding the "other" and his mind set. Whether it worked as intended would require postwar panels of military experts, historians and applied psychologists, not to exclude rubbernecking political scientists, to ascertain on a case by case basis. The complexity of apparently simple Cold War deterrence holds a cautionary tale for later students of history and politics. This point is emphasized by one expert Russian commentator who urges caution about a nuclear free world:

> In a nearly ideal world, neither the United States nor Russia would have needed large [nuclear] arsenals. In the world as it is, reduction to the minimum will only push into the foreground countries with small nuclear arsenals and elevate them to the status of world powers. All of that will stimulate a nuclear arms race on the global scale, a development certainly detrimental to security or stability.[15]

---

[15] Sergei Karaganov, "The START II Promotes Russian–American Cooperation and Rapprochement: Nuclear Free world is a dangerous concept that ought to be abandoned," *Rossiiskaya Gazeta*, April 23, 2010, in *Johnson's Russia List* 2010 – #81, April 23, 2010, davidjohnson@starpower.net.

Deterrence after the Cold War and into the present century is therefore not complicated compared to its mythical Cold War simplicity. It was always complicated and will remain so. If deterrence is ever oversimplified, it will work no longer. Like double agents and Swedish films, deterrence demands a certain mystique and ineffability for its realization in practice. It is software more than hardware, and it is more politics and strategy than it is tactics.[16] The essence of deterrence in the new world order is not that it is no longer pertinent to the challenges faced by the UNITED STATES and allied states. It is instead for them to comprehend the political landscape of the Middle East, South Asia and other theaters of military action that might fast forward challenges that were once on the periphery of concern for policy makers.

In addition, deterrence, like defense, lives with advanced technology but is not "about" technology per se. Cold War theorizing was sometimes handicapped by technological fetishism, and similar risks bestride post-Cold War thinking. An information-age RMA (Revolution in Military Affairs) may privilege the United States and other states with head starts in the development and deployment of systems for C4ISTAR (command, control, communications, computers, intelligence, surveillance, targeting and reconnaissance), long range precision strike, stealth and other by-products of the information, communication and electronics revolutions applied to warfare.[17] However, evident U.S. superiority in advanced technology conventional warfare will not make nuclear weapons superfluous. Instead, states outgunned by America's shock and awe machine will turn to asymmetrical approaches for deterrence of, defense against, or attack on U.S. interests – and these approaches might include threat or first use of WMD.

The United States and its allies must now find the means for deterrence of fighting networks in addition to state hierarchies, or, if networks are truly beyond deterrence, of defeating them.[18] However, since the supply of potential recruits for terrorist cells, including some that might be interested in WMD, is theoretically endless, deterrence by denial cannot rest on a futile strategy of attrition (viz. Petraeus' warning). Deterrence by credible threat of retaliation is also needed in the repertoire of states' responses to terrorist threats – but

---

[16] For insights to this effect, in the context of an important case study, see Amitai Etzioni, "Can a Nuclear-Armed Iran Be Deterred?," *Military Review*, May–June 2010, pp. 117–125.

[17] For perspective on the information-led RMA and its implications, see Lawrence Freedman, *The Transformation of Strategic Affairs* (London: International Institute of Strategic Studies, Adelphi Paper 379, and Routledge Publishers, 2006), esp. pp. 11–26, and Gray, *Strategy for Chaos: Revolutions in Military Affairs and the Evidence of History* (London: Frank Cass, 2002), for more comprehensive historical assessment of the RMA concept.

[18] This important point is argued in John Arquilla, *Worst Enemy: The Reluctant Transformation of the American Military* (Chicago, Ill.: Ivan R. Dee, 2008), passim.

retaliation that is lethal and focused for the purpose of disrupting networks, not destroying societies. In all likelihood, this excludes nuclear deterrence, *except* for the task of deterring nuclear weapons states in cahoots with terrorists.

Relevant or irrelevant utopias? abolition in whole or part?

Experience suggests that the objectives of nuclear arms limitation and confidence building are more within the reach of governments than the objective of nuclear abolition, however desirable it might prove to be. Nuclear abolition would not, for reasons discussed earlier, obviate the need for deterrence. In an abolition regime, conventional deterrence would be combined with nuclear forbearance so long as states regarded this situation as politically and militarily acceptable. When, on the other hand, states lost faith in a system based on nuclear abstinence and returned to bomb making and deployment, they could then redraw upon the storehouse of knowledge accumulated during the first and second nuclear ages. A third nuclear age, of post-abolition disillusionment replaced by nuclear realpolitik, might follow: and it might be more dangerous than either of the first two, as states with nuclear knowledge but denied arsenals now raced to get systems promptly deployed for deterrence or coercion missions.

This future, of post-abolition nuclear deterrence, is not here yet. What is on offer is a mosaic of: (1), U.S. and Russian nuclear forces, downsized from their high Cold War sizes but periodically modernized and figuratively, if not literally, pointed at one another; (2), second tier P-5 nuclear forces held by the United Kingdom, China and France, also undergoing modernization or planning for it; (3), the nuclear forces of India and Pakistan, officially acknowledged by their offsetting tests in 1998;[19] (4), the nondeclaratory, but de facto nuclear weapons state status of Israel; (4), the suspected aspirations of Iran for nuclear weapons;[20] (5), the apparent nuclear arsenal now available to North Korea,

---

[19] Important studies include: Alexander H. Rothman and Lawrence J. Korb, "Pakistan doubles its nuclear arsenal: Is it time to start worrying?," *Bulletin of the Atomic Scientists*, February 11, 2011, http://www.thebulletin.org/node/8607, downloaded February 15, 2011; Henry D. Sokolski, ed., *Pakistan's Nuclear Future: Worries Beyond War* (Carlisle, Pa.: Strategic Studies Institute, U.S. Army War College, January 2008); and, D.R. SarDesai and Raju G.C. Thomas, eds., *Nuclear India in the Twenty-First Century* (New York: Palgrave-Macmillan, 2002), esp. Ch. 5.

[20] For details on Iran's progress toward the capability for nuclear weaponization, see the Webpage for *Iran Watch*, "Iran's Nuclear Timetable," updated December 2, 2010, and regularly. http://www.iranwatch.org/ourpubs/articles/iranucleartimetable.html. Additional technical information, and informed discussion of U.S. and international community options for containing Iranian nuclear

absent diplomatic reversal of this situation;[21] (6), the nuclear ambivalents or undecideds among states with perceived security dilemmas, uncertain about the viability of the nonproliferation regime; and, (7), nonstate actors, including terrorists, who aspire to acquire fissile materials, assembled nuclear weapons, or other WMD for purposes ranging from the explicitly political to the apocalyptic. Of additional importance for nonproliferation historians, but not included here, are states that reversed a course of action initially plotted toward nuclear weaponization.

This picture is not an entirely depressing one. Nuclear weapons have spread less rapidly than Cold War pessimists had forecast.[22] As early as the 1960s, United States and other experts and politicians were warning that as many as thirty countries might have nuclear forces within decades. Why was there a slow pub crawl into the nuclear club instead of a mad rush? Many reasons present themselves, including the daunting obstacles facing a state that would construct a believable nuclear force, instead of a Potemkin village. Financial resources, military and scientific knowledge, and a civil–military relationship that can provide for both positive and negative control of nuclear forces are all necessities - thus limiting the membership of the select club of nuclear weapons states with survivable, flexible and controllable forces to a small number.

On the other hand, the globalization of information and communications now spreads news reports and images across the planet with dramatic suddenness. Among these images are those of states seeking to project strength for the benefit of friends, enemies and observers. Nuclear weapons can appeal to states with time-urgent security dilemmas that cannot be overcome by means of conventional deterrence, or defense, alone (e.g., Pakistan). Nuclear forces can also appeal to regional powers with large ambitions who seek to deter the United States or other outsiders from expeditionary interventions, including efforts at regime change (e.g., Iraq under Saddam Hussein before

---

weaponization, appear in David Albright and Jacqueline Shire with Paul Brannan and Andrea Scheel, *Nuclear Iran: Not Inevitable* (Washington, D.C.: Institute for Science and International Security, January 21, 2009).

[21] North Korea is a moveable feast. See Choe Sang-Hun, "North Korea Says It Has 'Weaponized' Plutonium," *International Herald Tribune*, January 17, 2009, http://www.iht.com/articles/2009/01/17/news/norkor.1-409776.php. Obama administration officials have expressed concerns about the possibility of a leadership crisis in North Korea. See "Clinton: U.S. fears N. Korea leadership crisis," *Associated Press*, February 19, 2009, http://www.msnbc.msn.com/id/29272217/html. Expert assessment appears in Andrew Scobell and John M. Sanford, *North Korea's Military Threat: Pyongyang's Conventional Forces, Weapons of Mass Destruction, and Ballistic Missiles* (Carlisle, Pa.: U.S. Army War College, Strategic Studies Institute, April 2007).

[22] Joseph Cirincione, *Bomb Scare: The History and Future of Nuclear Weapons* (New York: Columbia University Press, 2007), p. 125 and passim.

1991, or Iran and North Korea now). If nuclear weapons are now the preferred counters for the weaker against the stronger powers, the appeal of Global Zero becomes asymmetrical. As Ariel Levite has noted:

> Suspicion already abounds in the international community that the mighty powers are supporting Global Zero only in order to deprive or deny their otherwise asymmetric adversaries the unmatched nuclear equalizers they already possess (North Korea), are acquiring (Iran), or might elect to pursue.[23]

It should not be assumed, though, that only revisionist state or nonstate actors, who seek to overthrow the existing order or to disrupt international stability otherwise defined, retain an interest in nuclear weapons. No country included among the Permanent Members of the UN Security Council, who also constitute the NPT-approved original nuclear weapons states, has offered to unilaterally disarm and dismantle its nuclear arsenal. Despite the inclusion in the NPT of an expectation that the current nuclear weapons states would follow through with their own eventual disarmament, the appeal of nuclear weapons has outweighed the temptation to take risks for nuclear peace. This apparent hypocrisy about disarmament on the part of the P-5 has not gone unnoticed, nor unremarked, by prominent nonsignatories of the NPT and others. According to Russian nuclear arms control expert Sergei Karaganov:

> Denouncement of nuclear weapons and wishes to see them eliminated are moral. It is necessary to remember, however, that accomplishment of this particular objective is only possible and desirable if and when Man himself changed. Nuclear Zero champions apparently believe that Man is going to change. Sorry, I'm not one of them.[24]

Dr. Karaganov's "first image" skepticism is supported by some concerns about "third image" international systems theory as it might apply to nuclear danger. The Cold War was the justification for the United States, Russia, Britain, France and China to maintain nuclear weapons, but what now? The end of the Cold War and the demise of the Soviet Union have not changed the essence of the international political system – these events have merely rearranged the deck chairs on the *Titanic*. International systems, despite the insistence of some international relations theorists, do not dictate to states, and especially to the stronger states within the system, their primary decisions about ends, ways

---

[23] Levite, "Global Zero: An Israeli Vision of Realistic Idealism," p. 167.

[24] Karaganov, "The START II Promotes Russian–American Cooperation and Rapprochment."

and means. The international system is subsystem dominant. Powerful actors shape, and reshape, the international order, unless and until resistant national or international forces rise to balance against them.

No disrespect for international systems theory is intended by the preceding discussion. The international system matters a great deal as a level of analysis for theoretical explanation and prediction in world politics.[25] There are behavioral regularities among nations that are created by their systematic interactions, and these include important regimes for international law, arms control and other matters that cross state lines. Indeed, in certain aspects, finance and technology globalization create a more important "system" effect than hitherto, washing across national borders to impact directly upon consumers and governments. In some ways, we are indeed in a borderless world.

Thus the objection here is not to the *acknowledgment* of the importance of international systems. Danger lies, instead, in the *attribution* to systems of powers that they cannot exercise and basing expectations for peace on that attribution. For example: it is to Jonathan Schell's credit that he acknowledges the interactive relationship between nuclear abolition, which he urges, and world government.[26] Without world government, nuclear abolition lacks an enforcement capability that portends a short history. But, as Schell would doubtless acknowledge, getting to world government is no easier, and arguably harder, than getting to "nuclear zero" in the numbers of deployed or other nuclear weapons. States might give up their nuclear weapons but would doubtless cling to their sovereignty with even more tenacity. For a state to give up its sovereignty is to give up its identity and that of its people. Herein lies the rub.

The post-Cold War international order has been marked primarily by internal wars within states, as opposed to major interstate wars (with notable exceptions, such as the United States and allied wars against Iraq in 1991 and 2003). Many of these internal or substate wars have been fought over issues of identity (who are we?) in addition to the more traditional stakes of interstate wars (territory, economic resources, military power, dynastic succession). Wars of identity, including ethnic, religious, tribal and clan affiliations (often

---

[25] Competing theories of international politics, as Jack Snyder points out, perform various functions: they provide filters for interpreting complicated reality; they reveal the assumptions behind foreign and security policy debates; and, third, alternative theories help to check one another by identifying weaknesses in arguments. See Jack Snyder, "One World, Rival Theories," Ch. 1 in Karen A. Mingst and Jack L. Snyder, eds., *Essential Readings in World Politics*, Third Edition (New York: W. W. Norton, 2008), pp. 4–11. Equally important is the recognition that general theorists of world politics are "ever willing to pass beyond Clausewitz's 'culminating point of victory.'" Gray, *The Second Nuclear Age*, p. 23.

[26] Jonathan Schell, *The Seventh Decade: The New Shape of Nuclear Danger*. (New York: Henry Holt and Co., 2007).

mixed together in various ways), are not only about the primordial sources of tensions between disparate cultures trapped within the same polity. They are also about which culture or "nation" will rule by means of legitimated, or other, state power. From Bosnia to Bishkek, the paradoxical impact of the flattening economic and technological world is that sensitivities about origins, identity and personal meaning are more inflammatory and more "political." Perhaps this is not entirely paradoxical: perhaps, in an ever more globalizing world, particularistic affiliations seem in peril, and therefore people and leaders cling to them more tenaciously.[27]

Regardless of the impact of globalization on politics by way of cultural anthropology, the issue of empowered disparate ethno–national and religious groups, each seeking ownership of its own state, is pertinent to the quest for nuclear abolition. If state sovereignty means that each polity will be more vulnerable to internal dissatisfaction or revolt based on primordial values and affinities, then nuclear weapons may appeal more, instead of less, as symbols of sovereignty "above the fray" for national leaders. In 1998, the nuclear tests in India and Pakistan were followed by finger pointing and head shaking in Western foreign offices. In Indian and Pakistani villages, people reportedly danced in the streets. The appeal of nuclear weapons as a symbol of sovereignty is not limited to India and Pakistan, for obvious reasons. The symbolism is based upon destructive potential second to none among candidate weapons for mass destruction. That potential for nuclear destruction, attendant to even a "small" nuclear war, gets attention. The first nuclear weapon fired in anger since Nagasaki will grab the headlines and the attention of governments like no other event since 9/11.[28]

The objective of nuclear abolition, therefore, collides with the aspiration on the part of more peoples, including Palestinians, Kurds and others located in dangerous neighborhoods, for their own territorial state. Sovereignty is prized more, not less, in the new world order than hitherto. Against this, international systems theorists and policy analysts can make a reasonable argument based on the "tragedy of the commons." If nuclear "beggar thy neighbor" policies result in an unmanageable or unlimited spread of nuclear weapons, the entire

---

[27] Implications of a more technologically globalized, but politically fragmented, world include the priority of "culture" and "civilization" as axes of conflict. See Richard H. Shultz, Jr. and Andrea J. Dew, *Insurgents, Terrorists, and Militias: The Warriors of Contemporary Combat* (New York: Columbia University Press, 2006), esp. Ch. 1–3, and Samuel P. Huntington, *The Clash of Civilizations and the Remaking of World Order* (New York: Touchstone Books, 1998), esp. Chs. 1–2 and 10–11. Among studies of the military implications of these trends for the U.S. armed forces, see Robert D. Kaplan, *Imperial Grunts: On the Ground with the American Military from Mongolia to the Philippines to Iraq and Beyond* (New York: Vintage Books, 2005).

[28] George H. Quester, *Nuclear First Strike: Consequences of a Broken Taboo* (Baltimore, Md.: Johns Hopkins University Press, 2006), pp. 24–52 offers possible scenarios.

system may reach a point of punctuated equilibrium after which outbreaks of nuclear war become, if not routine, at least no longer "unthinkable." A worst case scenario is that nuclear weapons become included as tactical options for use as well as for deterrence, along with other instruments of military power (much as U.S. nuclear artillery shells and land mines were deployed in Europe during the Cold War, or their Soviet counterparts).

Here the systems theorists are correct, but their assumed cure for the disease of hyper-sovereignty combined with nuclear weapons involves a bypass through world government or, more realistically, a stronger nonproliferation regime. The former is beyond political attainability by all indications of past practice; the latter is not. Because world government is not a plausible path to nuclear abolition, nuclear limitation or regulation by means of interstate agreement is the next best hope. It must be said, first, that nuclear limitation is not uncontroversial. Some states benefit from the existing balance of nuclear power, and they can be expected to defend the prevailing order. However, the majority of world governments and even nuclear weapons states ought to confront the prospect of uncontrolled proliferation with dread, and there is room for improvement in the nonproliferation regime that would leave sovereignty intact, discard superfluous nuclear weapons, and contribute to a reduction in the nonlinearity, and potential chaos, attendant to loose nukes and new nuclear weapons states.

Toward that end, we have no quibble with the agenda set by the Obama administration for nuclear limitation: (1), further reductions in strategic nuclear weapons and delivery systems as between the Russians and the Americans, under a post-START I and post-SORT regime;[29] (2), U.S. and other outliers' support for, and ratification of, the Comprehensive Test Ban Treaty (CTBT);[30] (3), support for widespread ratification and subsequent verification of the Fissile Materials Cut-off Treaty; (4), reaffirmation and strengthening of the

---

[29] Expert commentary appears in Alexei Arbatov and Rose Gottemoeller, "New Presidents, New Agreements? Advancing U.S.–Russian Strategic Arms Control," *Arms Control Today*, July/August 2008, http://www.armscontrol.org/act/2008_07-08/CoverStory.asp. See also: Stephen J. Blank, *Russia and Arms Control: Are There Opportunities for the Obama Administration?* (Carlisle, Pa.: Strategic Studies Institute, U.S. Army War College, March 2009); Pavel Podvig, "Russia's new arms development," *Bulletin of the Atomic Scientists*, January 16, 2009, http://thebulletin.org/web-edition/columnists/pavel-podvig/russias-new-arms-development; and Steven Pifer, "Beyond START: Negotiating the Next Step in U.S. and Russian Strategic Nuclear Arms Reductions," Brookings Institution, www.brookings.edu, in *Johnson's Russia List* 2009 – #88, May 12, 2009, davidjohnson@starpower.net.

[30] The case for this is presented in Tom Z. Collina with Daryl G. Kimball, *Now More Than Ever: The Case for the Comprehensive Nuclear Test Ban Treaty* (Washington, D.C.: Arms Control Association, Briefing Book, February 2010). Prudent caution, with regard to possibly unintended consequences from CTBT ratification, is urged in Sokolski, "Moving Toward Zero and Armageddon?," p. 78.

Non-Proliferation Treaty (NPT) in the 2010 review conference; and, (5), backing for this and other measures intended to defend the "Allison line" (after scholar and security analyst Graham Allison) on proliferation (no loose nukes, no new nascent nukes, and no new nuclear weapons states) for as long as possible.[31]

These measures of nuclear limitation and strategic prudence will not constitute an anti-nuclear revolution. Nuclear limitation is not the same thing as nuclear abolition, or even nuclear marginalization. Far from having been marginalized, nuclear weapons remain important in the twenty-first century for those who hold them, for those who aspire to, and for those who are affected by both existing and aspirational nuclear weapons states. Living with the bomb is admittedly living below the best of all possible worlds, but it is also possible to live above the worst: bombs aplenty, in undisciplined and angry hands, crashing into deterrence and turning social values into depleted uranium.

In the "real world" – alternatives and options

If nuclear abolition is unattainable, at least immediately, it does not necessarily follow that nuclear arms limitation is undeserving of accomplishment.[32] In the case of the United States and Russia in their search for a limitation and reductions regime to replace START I, options were bounded by security commitments, by past practice, and by the willingness of governments and their negotiators to bargain toward a meaningful outcome. The resulting New START treaty achieved only incremental reductions in the two states' numbers of operationally deployed weapons, compared to those already committed in the SORT agreement, but New START also foreshadowed further cooperation in nuclear security issues as between Moscow and Washington.[33] The "rationality" of arms control negotiations does not lie in the numbers of nuclear weapons and long range launch vehicles, but in the assumed political

---

[31] Graham Allison, *Nuclear Terrorism: The Ultimate Preventable Catastrophe* (New York: Times Books – Henry Holt and Co., 2004). For additional expert analysis on this topic, see Brian Michael Jenkins, *Will Terrorists Go Nuclear?* (New York: Prometheus Books, 2008), esp. pp. 277–291 on the applicability of deterrence to contemporary and prospective threats of terrorism.

[32] This point is argued persuasively in George Perkovich, *The Obama Nuclear Agenda One Year After Prague* (Washington, D.C.: Carnegie Endowment for International Peace, March 31, 2010), and Daryl G. Kimball, "Next Steps on New START," Arms Control Today, April 2010, http://www.armscontrol.org, downloaded April 1, 2010.

[33] *Treaty between the United States of America and the Russian Federation on Measures for the Further Reduction and Limitation of Strategic Offensive Arms* (Washington, D.C.: U.S. Department of State, April 8, 2010), http://www.state.gov/documents/organization/140035.pdf.

rationale for each state's deployments. In the present world, Russia and the United States deploy long range nuclear weapons not only (or mainly) in view of their two sided relationship, but also in view of their global responsibilities and interests.

That having been said, Russia does regard the appearance of nuclear-strategic parity with the United States in long range weapons as a visible symbol of its entitlement to international influence and respect. This perception of U.S.–Russian nuclear-strategic parity was one of Russia's primary objectives in negotiating the New START agreement signed by Presidents Obama and Medvedev in April, 2010. In addition, Russia's nuclear forces, including its forces of intercontinental range, compensate across the deterrent spectrum for the deficiencies in its conventional military. The latter is about to undergo a serious modernization and reform that will transform it from a mass mobilization conscript force, on the old Soviet model, to one built mostly on voluntary service and with smaller, more mobile and high-tech equipped troops. But this military reform will take years, if accomplished at all; meanwhile, nuclear forces of various ranges will have to substitute for the lacunae in Russian conventional forces, for better or worse. This situation increases Russia's dependency on nuclear weapons, not only for deterrence of nuclear attack, but also for the deterrence of large scale conventional war posing a vital threat to the Russian state.[34]

The Obama administration announced in September, 2009 that it would scrap the Bush plan for missile defenses in Europe. Instead, the United States would deploy a reconfigured system that was focused more on intercepting Iranian short or medium range missiles, as opposed to the longer range ballistic missiles that Iran might eventually develop.[35] According to U.S. Secretary of Defense Robert Gates, the new plan would actually deploy missile defenses against Iranian threats seven years ahead of the Bush plan, and some land based interceptor missiles would be deployed in Europe, possibly in Poland.[36] Anticipating some negative reactions to the plan from Europe and the American home front, Gates added that the new BMD configuration "provides a better missile defense capability" for Europe and U.S. forces

---

[34] For Russia's current military doctrine, see: Text, "The Military Doctrine of the Russian Federation," www.Kremlin.ru, February 5, 2010, in *Johnson's Russia List* 2010 – #35, February 19, 2010, davidjohnson@starpower.net. On the nuclear implications of Russia's military doctrine, see Nikolai Sokov, "The New, 2010 Russian Military Doctrine: The Nuclear Angle," Center for Nonproliferation Studies, Monterey Institute of International Studies, February 5, 2010, http://cns.miis.edu/stories/100205_russian_nuclear_doctrine.htm.

[35] Peter Baker, "White House to Scrap Bush's Approach to Missile Shield," *New York Times*, September 17, 2009, http://www.nytimes.com/2009/09/18/world/europe/18shield.html.

[36] Ibid.

located there than the program he had recommended almost three years earlier (as President George W. Bush's Secretary of Defense).[37]

Missile defense technologies compared to offenses are immature and cannot transcend deterrence based on offensive retaliation, let alone supersede the nuclear revolution. But missile defenses, in the post-New START political context as between the United States and Russia, can provide an arena for political security cooperation and reassurance against Russia's worst case fears of possible nullification of its nuclear deterrent. The possibility of using U.S. and NATO missile defenses as instruments for cooperative security with Russia has been raised by NATO Secretary General Anders Fogh Rasmussen. In a commentary on the international security climate and nuclear challenges in the aftermath of the U.S.–Russian New START agreement, Rasmussen noted:

> We need a missile defense system that includes not just all NATO countries, but Russia, too. The more that missile defense is seen as a shared security roof built, supported and operated together that protects us all, the more people from Vancouver to Vladivostok will know that they are part of one community. Such a security roof would be a strong political symbol that Russia is fully part of the Euro-Atlantic family, sharing the costs and benefits.[38]

Rasmussen acknowledges that important military and technical obstacles stand in the way of NATO–Russian missile defense collaboration: the need to share intelligence, to make systems interoperable, to link sensitive technologies, and so forth.[39] But his vision of a possible Great Northern missile shield against possible sources of threat from outside of Europe (read: Iran for now, but who knows about later) is an agile deconstruction of foreign policy "Oldthink." It is also a javelin pointed directly at the cuirassiers of politicians in NATO and in Russia who cling to the mantras of the Cold War: that defenses are necessarily destabilizing; that the United States (or Russia) still seeks a nuclear first strike capability against the other leading nuclear state; and, that militaries in both countries can continue to plan against improbable, not to say fantastic, contingencies of nuclear Cold War redux, instead of collaborating in the face of new, and imminent, perils. It would be preferable not to wait for the outbreak of a nuclear arms race in the Middle East, or a successful act of nuclear terrorism, before pulling Old Soviets and unreconstructed Neocons across the River Styx.

---

[37] Ibid.

[38] Anders Fogh Rasmussen, "NATO's Common European Roof," *Moscow Times*, April 22, 2010, in *Johnson's Russia List* 2010 – #83, April 29, 2010, davidjohnson@starpower.net.

[39] Ibid.

Missile defenses are only one kind of advanced, non-nuclear technology calling into question past approaches to arms control, disarmament and nonproliferation. Another example of possibly paradigm-pushing technologies are those for long range, prompt global strike. U.S. plans to develop and deploy such weapons on formerly nuclear assigned launchers raised concerns on the part of Russian interlocutors at the New START negotiations. Russians fear that American PGS weapons fired from intercontinental launchers could be mistaken for a nuclear first strike. U.S. plans for PGS weapons involve programmed launch trajectories, and apparent inspection rights for foreign countries, to provide reassurance on this point. The objective of U.S. PGS systems would be to provide a conventional alternative to nuclear systems for timely attacks against important, but elusive, threats – including terrorists preparing an imminent attack with weapons of mass destruction, including nuclear ones.[40] On the other hand, some experts are skeptical of the need for, or the feasibility of, PGS weapons. Nuclear policy expert Joseph Cirincione, for example, contends:

> Using intercontinental ballistic missiles to hurl conventional weapons at caves is a truly bad idea. It would use technology that doesn't work for a capability the United States doesn't need at a cost it can't afford.[41]

Russia's suspicions about U.S. intentions in conventionalizing some nuclear capable launchers are perhaps understandable, but mistaken. Both states should welcome an opportunity to download nuclear tipped weapons and replace them with conventionally armed ones, provided appropriate transparency is maintained. New START and post-New START negotiations are about doing away with *nuclear* weapons, not weapons per se. A conventionally armed weapon fired against a distant target, if launched by mistaken decision or in response to faulty warning, allows for less collateral damage and a reduced possibility of starting World War III than does a mistaken or accidental strike by nuclear weapons. In the latter case, the psychological impact alone of the first nuclear weapon fired in anger since Nagasaki, apart from the actual physical damage to life and property, will be a transformational event in world politics – to use an already overworked term. Instead of running away from partial conventionalization of their long range launchers, the United States and Russia might collaborate on shared reconnaissance and global strike capabilities against common enemies, for example, terrorists.

---

[40] Craig Whitlock, "U.S. developing new non-nuclear missiles," *Washington Post*, April 8, 2010, http://www.msnbc.msn.com/id/36253190/ns/us_news-washington_post/print/1/.

[41] Joseph Cirincione, "Global Strikeout," *Foreign Policy*, April 23, 2010, http://www.foreignpolicy.com/articles/2010/04/23/global_strikeout. The same mission can be accomplished by drone aircraft within or near the theater of operations, according to the author.

Conclusions

Nuclear abolition can be desirable in principle but devilishly difficult to accomplish in practice. Large military and political speed bumps await efforts to get to nuclear zero, and political obstacles are the more obstinate. For better or worse, nuclear weapons are symbols of national sovereignty and international power that confer prestige in the minds of some, support "extended" deterrence for friends and allies, and backstop coercive diplomacy across a variety of issues. On the other hand, nuclear weapons, thought to be system stabilizers during the Cold War, are potential contributors to system disruption in the twenty-first century. Nuclear weapons in the hands of terrorists are a nightmare perhaps beyond deterrence. And the appeal of nuclear weapons for additional countries in Asia or in the Middle East is a mistaken judgment that bases strategy on disruptive technology, for those states, and a potential entrapment for others, including the United States and Russia.

Even if nuclear abolition is impossible to accomplish, nuclear limitation and risk reduction are achievable. The United States and Russia have agreed to reduce their numbers of deployed strategic nuclear weapons below SORT-prescribed levels in the New START regime. A post-New START agreement following on the success of the 2010 pact might reduce the numbers of Russian and American operationally deployed long range nuclear weapons to a maximum of 1,000 for each. Permitting each state to maintain 1,000 deployable weapons would provide sufficient numbers of warheads for existing and foreseeable nuclear deterrence and reassurance missions, and within tolerable parameters of crisis stability.

U.S.–Russian nuclear force reductions by themselves can still leave at risk the nonproliferation regime, unless these bilateral accomplishments are supported by the multilateral agreements listed earlier in this discussion (section five), plus others.[42] If North Korea is permitted to maintain and Iran to achieve nuclear weapons status, stability in Asia and the Middle East is imminently imperiled. The preceding points are valid whether missile defense technology is robust or primitive relative to offenses. Unconstrained proliferation of nukes and ballistic missiles will leave defenses in a tail-chasing mode – defenses, like offenses, require for their success as instruments for deterrence a constrained proliferation regime.

---

[42] For an informative argument to this effect, see Sokolski, "Moving Toward Zero and Armageddon?"

5

After the loving: New START and beyond

Introduction

The Obama administration, following a protracted negotiating process with the U.S. Congress, equal in difficulty to the preceding negotiations with Russia, finally obtained ratification of the New START agreement on strategic nuclear arms reductions in December, 2010. How much does this accomplishment really matter, in politics, arms control and strategy? Advocates see New START as a door opener to further progress in nuclear arms reductions, nonproliferation and related issues among the U.S., Russia and other nuclear weapons states. Skeptics on the left and right of the political spectrum wonder whether New START is a passé example of Cold War redux, or evidence that arms control can succeed only when enabled by a permissive political climate that may not last.

Seeing New START in context requires both general and specific discussion – of general issues related to the Obama policy agenda with Russia and nuclear weapons and of specific consideration of whether the New START agreement once implemented will provide for stable deterrence at lower numbers of weapons. This discussion considers the prospects for strategic nuclear arms reductions as follows: (1), the more extensive Obama agenda for nuclear marginalization and, in extremis, nuclear abolition; (2), the responsibility of the United States and Russia to provide global leadership for nuclear nonproliferation; (3), the prospects for a reset in NATO–Russian as well as U.S.–Russian relations as related to issues of nuclear arms limitation; and, (4), across the preceding three discussions, the significance of missile defense as an issue for U.S.–Russian and NATO–Russian political and military cooperation.

Aiming high

U.S. President Barack Obama in 2009 called for international support for the objective of global nuclear abolition, even as he acknowledged that the job would take a long time to accomplish and meet with inevitable resistance.[1] From the outset, the Obama administration committed the United States to an ambitious agenda with respect to the reduction of global nuclear danger.[2] This agenda included: (1), the accomplishment of a New START agreement with Russia on the reduction of long range or "strategic" nuclear weapons, signed in April, 2010 by Presidents Obama and Dmitry Medvedev and ratified by the U.S. Congress in December, 2010;[3] (2), resubmission of the Comprehensive Test Ban Treaty (CTBT) that was signed by the Clinton administration, but rejected by the U.S. Senate for ratification in 1999; (3), a review and extension conference for the Nuclear Non-Proliferation Treaty (NPT) in May, 2010, in New York; (4), in line with post-START and NPT objectives, encouragement of other nuclear weapons states to reduce their numbers of deployed nuclear warheads and nuclear-capable launchers; and, (5), international efforts on the part of the IAEA (International Atomic Energy Agency) inspectors and various negotiating "contact groups" to disarm North Korea as a nuclear weapons state and to prevent Iran from joining the ranks of military nuclear powers.[4] This activist schedule of arms control and disarmament objectives is by no means the endgame for an ambitious American president. Nuclear arms reductions and nonproliferation

---

[1] Jennifer Loven, "Obama outlines sweeping goal of nuclear-free world," *Associated Press*, April 5, 2009, in *Johnson's Russia List* 2009 – #66, April 5–6, 2009, davidjohnson@starpower.net.

[2] See Remarks by Rose Gottemoeller, Assistant Secretary, Bureau of Verification, Compliance, and Implementation, U.S. Department of State, "The Long Road From Prague," Colonial Williamsburg, Virginia, August 14, 2009, http://www.state.gov/t/vci/rls/127958.htm.

[3] David E. Sanger, "After Treaty, Obama Nuclear Agenda Only Gets Harder," *New York Times*, December 21, 2010, http://www.nytimes.com/2010/12/22/us/politics/22assess.html. See also: Peter Baker, "Russia and U.S. Sign Nuclear Arms Reduction Pact," *New York Times*, April 8, 2010, http://www.nytimes.com/2010/04/09/world/europe/09prexy.html, and "Obama, Medvedev sign historic arms deal," *Associated Press*, April 8, 2010, http://www.msnbc.msn.com/id/36254613/ns/politics-white_house/print. See also, for expert analysis and projections: Pavel Podvig, "New START Treaty in numbers," from his blog, *Russian strategic nuclear forces*, April 9, 2010, http://russianforces.org/blog/2010/03/new_start_treaty_in_numbers.shtml.

[4] For expert assessment of the Obama nuclear arms control agenda, especially the post-START and Comprehensive Test Ban Treaty ratifications, see Robert S. Norris, "The Senate and the START treaty," *Washington Times*, November 12, 2009, http://www.washingtontimes.com/news2009/nov/12/norris-the-senate-and-the-start-treaty/html. See also: Brent Scowcroft, Joseph Nye, Nicholas Burns and Strobe Talbott, "U.S., Russia must lead on arms control," *Politico.com*, October 13, 2009, http://dyn.politico.com/printstory.cfm?

are way stations on the road to the eventual abolition of nuclear weapons worldwide.[5]

The Obama administration also rebooted the George W. Bush plan for the European-based component of the U.S. global missile defense program. The Obama ballistic missile defense (BMD) plan emphasizes a phased, adaptive deployment of theater missile defenses keyed on regional threats against European or other allied targets.[6] The immediate concern on the part of U.S. and allied military planners is their assumption that Iran is bent on acquiring a nuclear weapons capability or, short of that, developing the technology to become a virtual nuclear weapons state (i.e., maintaining a civil nuclear infrastructure and sufficient quantities of weapons-grade material to support a prompt decision for weaponization). Russia has ambivalent objectives on this issue: Moscow prefers that Iran not develop or deploy nuclear weapons, but Russia is more reluctant than are the U.S. and other "P-5 plus one" countries to impose stronger sanctions against Tehran. Russia also has strong economic ties to Iran (as does China) and neither Moscow nor Beijing would vote for the use of military force to dissuade Iran from joining the ranks of nuclear weapons states.

U.S. European-based missile defenses are regarded as ambiguous political expressions and potential military threats to Russia, in part because they overlap with NATO enlargement and Russian military reform. Russia asserts priority interests and a diplomatic *droit de regard* over security decisions in former Soviet space, but Russia's conventional military forces are inferior to those of NATO. Russia is embarked upon a serious military reform, intended to transform its armed forces from a World War II style military based on mass

---

[5] The case for nuclear abolition is argued by four influential policymakers in George P. Shultz, William J. Perry, Henry A. Kissinger and Sam Nunn, "Toward a Nuclear-Free World," *Wall Street Journal*, January 15, 2008, p. A13.

[6] On current and prospective U.S. missile defense programs, see Unclassified Statement of Lt. Gen. Patrick J. O'Reilly, Director, Missile Defense Agency, before the House Armed Services Committee, Subcommittee on Strategic Forces, Regarding the *Fiscal Year 2011 Missile Defense Programs* (Washington, D.C.: House Armed Services Committee, U.S. House of Representatives, April 15, 2010). Assessments of the revised Obama missile defense plan include: George N. Lewis and Theodore A. Postol, "A Flawed and Dangerous U.S. Missile Defense Plan," *Arms Control Today*, May 2010, http://www.armscontrol.org/print/4244; George Friedman, "The BMD Decision and the Global System," Stratfor.com, September 21, 2009, in *Johnson's Russia List 2009 – #175*, September 22, 2009, davidjohnson@starpower.net; Alexander Golts, "Calling Moscow's Bluff on Missile Defense," *Moscow Times*, September 22, 2009, in *Johnson's Russia List 2009 – #175*, September 22, 2009, davidjohnson@starpower.net; Alexander L. Pikayev, "For the Benefit of All," *Moscow Times*, September 21, 2009, in *Johnson's Russia List 2009 – #174*, September 21, 2009, davidjohnson@starpower.net; and Strobe Talbott, "A better base for cutting nuclear weapons," *Financial Times*, September 21, 2009, in *Johnson's Russia List 2009 – #174*, September 21, 2009, davidjohnson@starpower.net.

conscription and trained for decisive shock battle on the plains of Europe. Instead, the Medvedev–Putin "tandem" political leadership, the Defense Minister and the Chief of the General Staff recognize that twenty-first century Russia needs a modern military that is light, mobile, based on voluntary enlistment more than conscription, and capable of conducting network-centric warfare against conventional and unconventional military opponents. This military transformation will take considerable time even if successful, and, meanwhile, Russia's nuclear forces will be relied upon for deterrence and dissuasion missions other than those that are strictly nuclear.

The Obama administration like its predecessor has sought to find common ground with Russia where possible while not giving away its basic or extended deterrence commitments. With regard to the military balance as between Russian and U.S. strategic nuclear weapons, this implies that the two states can negotiate bilateral reductions in weapons deployed on intercontinental launchers without placing into jeopardy deterrence or crisis stability for either side. Some Russian and even some American analysts have warned that, if Russia's nuclear modernization fails, the U.S. will end up with a potential nuclear first strike capability relative to Russia's weaker strategic nuclear "triad" of land and sea based intercontinental missiles and heavy bombers.

In this instance, analysts may find that their statistical calculations lead to scientific impasses and briefings that dead-end in political incredulity. The truth is that the survival of a U.S.–Russian nuclear arms race beyond the Cold War and the Soviet Union has as much to do with bureaucratic inertia, atavistic habits of strategic thinking, and political opportunism, as it does with geostrategic logic. To this argument, about the political absurdity of nuclear war between the U.S. and Russia, pessimists respond that Russia might be coerced diplomatically by a superior U.S. strategic nuclear force, even if no weapons were actually fired in anger. However, as noted above, neither the U.S. nor NATO requires nuclear weapons for the coercion of Russia, if need be (as seen from the Russian perspective, which is what deterrence is all about).

The New START and possible follow-on agreements for strategic nuclear arms reductions might also expedite the building of stronger international consensus in favor of tough sanctions and other measures to support nuclear nonproliferation. The signature cases of Iran and North Korea are tipping points for the nonproliferation regime built around the Nuclear Non-Proliferation Treaty (NPT) and other international agreements. The July, 2010 visit to the Korean DMZ by Secretary of State Clinton and Secretary of Defense Gates, announcing additional sanctions against North Korea for its recalcitrance, was both a warning and an opportunity for North Korea to resume six-party talks on denuclearization. Success may await the departure of Kim Jong-il as head of state or the defenestration of his regime, and even then, denuclearization of North Korea may require a broader agreement that also officially terminates the Korean War and agrees to treat the DPRK as a normal state.

Obama's vision of a world without nuclear weapons is unlikely to be realized in his lifetime, as he has acknowledged. On the other hand, governing is about practical choosing, and our choices are not limited to nuclear abolition or nuclear anarchy. Nuclear containment and partial disarmament by disaggregation of the motives and means for nuclear weapons acquisition or, even worse, nuclear first use, is a worthy set of accomplishments. Nuclear containment begins with U.S. and Russian nuclear arms reductions (including eventually the thorny problem of nonstrategic nuclear weapons), continues with the drawing of firm lines against additional proliferators, and may proceed into more transparency and accountability as proposed in the Fissile Materials Cutoff Treaty (FMCT), in ratification of the Comprehensive Test Ban Treaty (CTBT), and in agreed international restrictions on the protection and accountability for fissile materials and nuclear production facilities.[7]

Alas for nonproliferation optimists, these worthy objectives cannot be accomplished by international diplomacy alone. Behind the aide memoire of the negotiator must lie the power of military potential for coercion or, if necessary, use against the state or nonstate enemies of humanity and decency. A very important aspect of nuclear danger in the twenty-first century, unlike the first nuclear age, more or less concomitant with the Cold War, is that nuclear weapons politically speaking are no longer "mass" but "niche" forces. That is: instead of supporting threats of societal destruction over large national territories or continental areas, nuclear weapons might provide for more specific threats of "access denial" against otherwise superior conventional forces in particular regions.

As advanced militaries move further into information-based conventional warfare, those left behind will be tempted toward nuclear and other weapons of mass destruction (WMD) as equalizers and system disturbers. Although both Cold War American and Soviet leaders eventually agreed that a nuclear war "could never be won, and should never be fought," as President Reagan said, definitions of "winning" are distressingly culture bound, ideologically fallible, and rooted in domestic psychologies that would defy Freudian analysis. In addition, whether new nuclear weapons states will have militaries and forces that are under the secure and stable command and control of sensible

---

[7] Supporting arguments on this point appear in Alexander H. Rothman and Lawrence J. Korb, "Pakistan doubles its nuclear arsenal: Is it time to start worrying?," *Bulletin of the Atomic Scientists*, February 11, 2011, http://www.thebulletin.org/node/8607, downloaded February 15, 2011.

political leaders (sensible meaning, in this context, deriving political objectives from international system realities, as opposed to domestic political fantasies) remains to be seen.

The Obama national security strategy, like the George W. Bush strategy that preceded it, is pessimistic about the likelihood that the further spread of nuclear weapons, especially among rogue state or nonstate actors, can advance the causes of peace and security. This consensus among political and military experts goes back even further, as Cold War history shows. Theorists can contemplate the spread of nuclear weapons with equanimity, on the assumption that deterrence having survived the Cold War will last out another century or so. However, deterrence is a problematical construct, culturally dependent and motivationally based on "the threat that leaves something to chance," as Thomas Schelling explained it.[8] In other words, nuclear deterrence is the military equivalent of a hedge fund or a derivative: it can appear to have worked infallibly until the very moment that it fails spectacularly. A small nuclear war is a contradiction in terms, not only in military terms but also in the psychological and political impacts of the first nuclear weapon fired in anger since Nagasaki.

Leadership is required

New START if successful will carry forward progress into both vertical (reductions in the sizes of existing arsenals) and horizontal (limitations on the numbers of nuclear weapons states) disarmament. Neither Washington nor Moscow can avoid its responsibility for leadership in nuclear nonproliferation and disarmament. They must lead because they are the largest shareholders of the world's nuclear weapons, because they have the largest inventories of deployed and ready long range nuclear launchers, and because they have the longest history of managing nuclear operations without war. Failure on the part of the United States and Russia opens the door not only to nuclear weapons spread in the Middle East and Asia, but also to the possible first use of nuclear weapons in anger since Nagasaki – with all of its attendant consequences for world order, including the possible demise of the nonproliferation regime itself.[9]

---

[8] Thomas C. Schelling, *Arms and Influence* (New Haven: Yale University Press, 1966), p. 108 and p. 121, note 8.

[9] On this point, see: Michael Krepon, *Better Safe than Sorry: The Ironies of Living with the Bomb* (Stanford, Calif.: Stanford University Press, 2009); Joseph Cirincione, *Bomb Scare: The History and Future of Nuclear Weapons* (New York: Columbia University Press, 2007); Jonathan Schell, *The Seventh Decade: The New Shape of Nuclear Danger.* (New York: Henry Holt and Co., 2007); George H. Quester, *Nuclear First Strike: Consequences of a Broken Taboo* (Baltimore, Md.:

The history of the nuclear age is one of infinite regression: every "proliferator" including the first, the United States, was once a non-nuclear weapons state.[10] China was once considered a rogue nuclear weapons state, and some leaders in both the United States and Russia recommended preemptive attacks against China's fledgling nuclear capabilities. Israel has never officially acknowledged its nuclear weapons capability, but unofficially has let the world know that it is prepared, in extremis, to use the nuclear weapons that it officially does not have. India, notwithstanding its Gandhian traditions, became a nuclear weapons state in order to balance against China, and Pakistan became a nuclear weapons state in order to balance against India. As Billy Joel might say: "And so it goes."

For nuclear arms control and disarmament to succeed, current nuclear weapons states must follow through on their obligations under the Nuclear Non-Proliferation Treaty to reduce their own numbers of deployed and stored nuclear weapons. However, this process of cooperative detoxification from nuclear addiction will not be easy to accomplish. Nuclear weapons appeal to states for reasons of security (they feel threatened, or they wish to intimidate others), of prestige (membership in elite clubs always carries its own cachet), and of domestic politics (nukes can be symbols of national or cultural pride). In addition, all security dilemmas are not equal. A briefing on nuclear abolition might be received with more politeness in military staff colleges or think tanks in the United States or in Britain, compared to its probable reception in Islamabad, New Delhi or Pyongyang.

The preceding point is not meant to imply that newer nuclear powers are necessarily more irresponsible in the management of their nuclear arsenals than grizzled veterans of the nuclear age. Instead, the argument simply acknowledges that, with respect to the probability of nuclear first use, some heads of state and military chiefs of staff have more reason to sleep with one eye open than do others. Nations with contiguous borders may be able to inflict "strategic" consequences on one another by using weapons regarded by the Americans and Russians as tactical, that is, of shorter range and less destructiveness. Nuclear neighbors in Asia, and possibly nuclear neighbors in the Middle East, instruct the mind wonderfully about the risks inherent in offensive, compared to defensive, realism, under the shadow of nuclear deterrence – even if one has accepted the major premises of the realist paradigm for understanding international relations.[11]

Johns Hopkins University Press, 2006); and Graham Allison, *Nuclear Terrorism: The Ultimate Preventable Catastrophe* (New York: Times Books – Henry Holt and Co., 2004).

[10] A point made with special clarity by Schell, *The Seventh Decade*, passim.

[11] On the distinction between offensive and defensive realism, see John J. Mearsheimer, *The Tragedy of Great Power Politics* (New York: W. W. Norton, 2001), passim.

In short, virtuous behavior on the part of existing nuclear weapons states does not guarantee that non-nuclear states, including nuclear aspiring and nuclear threshold powers, will remain free of sin. Politics is the high priestess of historical indeterminacy. For example, while some reductions in U.S. and Russian strategic nuclear forces are obviously contributory to nonproliferation and disarmament, it is not self evident that reducing U.S. and Russian nuclear forces to "minimum deterrents" of several hundred weapons, let alone abolition of those forces, would be contributory to additional security or peace. "How much is enough" is a calculation, not only about the numbers of nuclear capable weapons and launchers that a state needs to deploy, but also about the state of play among actors' political intentions and proclivities for nuclear first use or prompt response during crises.

The unfortunate fact of strategic history is that for a peace to endure, some state or group of states must enforce that peace.[12] Even if one passes this buck of enforcement to the "international community," it still requires the diplomatic collaboration and concerted military action of the great powers in any particular international system. Peace is not self sustaining. Accordingly, the task of disciplining a twenty-first century international peace, with or without nuclear weapons, will fall to a relatively few well endowed major states with robust militaries and prodigious budgets, as well as states with regimes and peoples prepared to pay the prices of armed constabulary work. NATO's military commitment in Afghanistan at this time of writing, tasked with counterinsurgency and counterterror missions while engaged in armed nation building, makes the point. NATO is in Kabul and Kandahar because there is no other alliance or international body that will accept responsibility to deny future jihadists a safe haven for terrorist plotting – including the plotting of attacks with nuclear weapons.

In this problematical and indeterminate international context, missile defenses, presuming that they work at all, will not be panaceas that impose a technological defeat on offensive nuclear weapons. Instead, missile defenses may be part of a package of disincentives, including diplomacy, deterrence, containment, nuclear risk reduction and military strength, that permit the United States, Russia and the international community to hold back the potential floodtide of nuclear weapons spread.[13] Precedent is on the side of the "good guys." Nuclear weapons spread much more slowly among states in the first nuclear age (roughly coinciding with the Cold War) than many experts predicted and pessimists expected. History is open to management by the

---

[12] See Colin S. Gray, *War, Peace and International Relations: An Introduction to Strategic History* (New York: Routledge, 2007), esp. pp. 264–279 and passim.

[13] This argument is made by Krepon, *Better Safe than Sorry*, esp. pp. 174–212.

responsible powers on behalf of stability or change – for the better, or for the worse. For example, missile defenses might be part of a defense-protected build-down of offensive nuclear weapons, as insurance against sneaky breakouts from disarmament or arms control agreements, or as deterrents by virtue of their ability to deflect small attacks.[14]

Colin S. Gray has noted that "peace" has at least two principal meanings: that war is not taking place now; and, second, that war is unthinkable and impossible in the exigent circumstances of international or regional order.[15] A true security community only exists in the second situation, as in NATO Europe presently. However, it is also the case that the international institutions, including nonproliferation regimes and supporting technologies (perhaps for inspections *and* defenses), cannot carry the ball alone toward the objective of enduring peace. Shared cultural values and compatible, if not identical, readings of history are equally important as are institutions and mechanisms for dissuasion, deterrence and defense.[16] Soft power and hard power, both persuasion and kinetic capability, are co-conspirators in the construction of durable peace with fewer, or no, nuclear armed states.[17] In this regard, NATO's overtures to Russia coinciding with its 2010 summit and new strategic concept (see below) were timely and, perhaps, politically shrewd.

New politics or old?

If a truly post-Cold War politics, permissive of more expansive thinking about nuclear arms reductions, emerges as between the United States and Russia, then realistic thinking about their minimum essential requirements for strategic (or other) nuclear forces might proceed apace.[18] One might begin by asking

---

[14] For an example of a Cold War proposal of this type, see Freeman Dyson, *Weapons and Hope* (New York: Harper Colophon Books, 1985), pp. 272–285.

[15] Gray, War, Peace and International Relations, p. 278.

[16] Ibid.

[17] Robert O. Keohane and Joseph S. Nye, *Power and Interdependence* (Third Edition) (New York: Longman, 2001), esp. pp. 20–32.

[18] Complexities in resetting Russian–European relations, with emphasis on Russian–EU relations, are explored in Andrei Fedyashin, "Is Russia's bid for Europe 'without borders' just wishful thinking?," *RIA Novosti*, September 8, 2010, in *Johnson's Russia List* 2010 – #171, September 8, 2010, davidjohnson@starpower.net. See also: "Head of Russian think tank mulls prospects for cooperation with NATO," *Interfax*, September 9, 2010, in *Johnson's Russia List* 2010 – #173, September 10, 2010, davidjohnson@starpower.net, and "NATO stuck between past and future: Lavrov," in *www.russiatoday.com*, September 1, 2010, in *Johnson's Russia List* 2010 – #166, September 1, 2010, davidjohnson@starpower.net.

exactly how many targets must the United States be able to destroy in Russia, or Russia in the United States, and what kinds of targets are necessary?

Some twenty years after the end of the Cold War, it is past time for rethinking of nuclear war plans and the underlying concept of maximum deterrence as between Russia and America. Instead, the framework or context for further planning should be one of cooperative security, based on minimum deterrence and forces configured for "retaliation only" and delayed retaliation at that. The United States has gravitated from a Single Integrated Operations Plan (SIOP) for nuclear war to a cafeteria menu of plans for a variety of possible contingencies. Regardless, it would be prudent to assume that Russian targets provide a significant proportion of the ground zeros for the notional and START-accountable U.S. warheads deployed on intercontinental launchers, and an equally significant percentage of the U.S. warheads actually deployed now (around 2,200 in anticipation of fulfilling SORT requirements scheduled to come into force at the end of the year 2012, now superseded by New START ceilings of 1,550 operationally deployed strategic nuclear warheads and 700 deployed intercontinental launchers).

What to do with U.S. tactical nuclear weapons deployed under NATO aegis in Europe? A minimum number could be kept in storage, as is now done at several locations in NATO Europe, but this idea is growing less popular with the host countries as time passes.[19] And forward deployed nukes invite preemptive attack on their storage sites, causing collateral damage to European civilians as well as military personnel. NATO's expert task force, previewing the 2010 version of the alliance's new strategic concept, did not preclude the possibility of reductions in NATO and Russian nonstrategic nuclear weapons. But the task force anticipated that negotiations on both sides would be difficult and complicated, and that, meanwhile, NATO required deployed tactical nukes for deterrence of potential threats and for reassurance of interallied commitment.[20]

---

[19] For important insights on this problem, see Pavel Podvig, "What to do about tactical nuclear weapons," *Bulletin of the Atomic Scientists*, February 25, 2010, http://the bulletin.org , in *Johnson's Russia List* 2010 – #43, March 3, 2010, davidjohnson@starpower.net, and Jacob W. Kipp, "Russia's Tactical Nuclear Weapons and Eurasian Security," *Jamestown Foundation Eurasia Defense Monitor*, March 5, 2010, in *Johnson's Russia List* 2010 – #46, March 8, 2010, davidjohnson@starpower.net. For detailed information on U.S. tactical nuclear weapons deployed in Europe, see Hans M. Kristensen, *U.S. Nuclear Weapons in Europe: A Review of Post-Cold War Policy, Force Levels, and War Planning* (Washington, D.C.: Natural Resources Defense Council, February 2005). As of 2005, an estimated 480 weapons were deployed in Europe, all gravity bombs and versions of the tactical B61. Some 300 bombs were assigned for delivery by U.S. F-15E and F-16C/D aircraft deployed in Europe or rotated through American bases. The remaining 180 bombs were designated for delivery by the air forces of five European NATO countries: Belgium, Germany, Italy, the Netherlands, and Turkey (Ibid., pp. 8–11).

[20] *NATO 2020: Assured Security; Dynamic Engagement, Analysis and Recommendations of the Group of Experts on a New Strategic Concept for NATO* (Brussels: North Atlantic Treaty Organization, May 17, 2010).

On the other hand, prospects for agreement as between NATO and Russia on missile defense and other security issues appeared to improve in 2010, relative to past tensions. NATO's summit of heads of state and government in November, 2010 endorsed a new strategic concept that called for a "true strategic partnership" between the alliance and Russia, asserted that NATO "poses no threat to Russia," and included specific mention of European missile defenses among other items for enhanced consultation and cooperative security.[21] On the other hand, the same strategic concept paper asserted that NATO will continue to "maintain an appropriate mix of nuclear and conventional forces" and "further develop NATO's capacity to defend against the threat of chemical, biological, radiological and nuclear weapons of mass destruction."[22]

The political undercurrent to these and other negotiations between NATO and Russia over tactical nukes and anti-missile defenses was marked by a larger disagreement at the level of strategic constructivism. Is twenty-first century Europe to be conceptualized as a divisible entity as between Russian-dominated and NATO spheres of influence, or as a common security space within which major states and alliances can devise cooperative solutions to problems, including regional missile threats, terrorism, energy security, environmental degradation, and other issues that call for multinational collaboration? The answer is not obvious, since, as a former Speaker of the U.S. House of Representatives once famously said, all politics are local. The expectations and priorities of presidents, prime ministers and parliaments will be the proving grounds on which concepts of divided or inclusive Europe are road tested.

Are proposals for post-New START reductions in U.S. and Russian strategic and tactical nukes too optimistic, and perhaps, utopian? Possibly, although a case could be made that it is also realistic. Nuclear abolition, if it ever happens, cannot be accomplished in the fashion of Chairman Mao's proposed economic Great Leap Forward. Politics and policy move in increments, especially within multinational alliances (like NATO) or as between sometime-deterred and sometime-reassured cooperative security partners (such as the Americans and Russians). Would the French, Chinese or Israelis make reductions in their arsenals proportionate to those committed by the United States and Russia, under a next-after-New START regime? Not necessarily – but a viable arms reduction and nonproliferation regime might not require proportional reductions among all declared and acknowledged nuclear weapons states.

---

[21] *Active Engagement, Modern Defence*, "Strategic Concept for the Defence and Security of The Members of the North Atlantic Treaty Organization," Adopted by Heads of State and Government in Lisbon, http://www.nato.int/cps/en/natolive/official_texts_68580.htm. Downloaded January 3, 2011.

[22] Ibid.

Do the preceding arguments provide reasons for U.S. or Russian complacency about the stability of nuclear deterrence under New START or other regimes? Complacency would be a mistake, for several reasons. First, deterrence is unreliable in practice, however elegant the assertion of deterrence theories. Deterrence is an art, not a science, and highly conditional on the particular circumstances of the relationship between the immediate deterrer and deterree.[23] Second, nuclear deterrence is overweighed with moral ambiguity. One advances one's policy agenda by making a credible threat that one would, unless desperate or insane, not wish to carry out. Yet the credibility of nuclear threats requires that the decision maker possess both the capability and the will to execute his threat. Even under circumstances of extreme duress, including warning of presumed enemy attack, moral persons will find themselves with divided psyches and frustrated search engines for alternatives to nuclear war. Third, we have no experience with actual, as opposed to virtual, nuclear wars. No one really knows how governments, militaries and societies will react, even under the conditions of a "small" nuclear war between regional, as opposed to global, nuclear powers. This makes prognostication of "how much is enough" for deterrence difficult, if not impossible. As Keith Payne has noted:

No one can identify *the* minimum number of nuclear weapons that is adequate to make deterrence "work" reliably as desired because there is *no such magic single number*. The requirements for deterrence vary across time, circumstance, and opponent and there is no methodology for knowing with such precision what, if any, level of capability is sure to provide the desired deterrent effect.[24]

Payne's warning about a willingness to depend on simple metrics for complicated deterrence problems is part and parcel of the debate about nuclear proliferation. Some contend that the slow spread of nuclear weapons among state actors does not necessarily bode ill for international stability or security. This optimism is based in part on the assumption that nuclear deterrence should work about as reliably in the future as it has in the past.[25] Some arguments for proliferation permissiveness draw upon international

---

[23] Among prominent strategic thinkers who have made this point, see Lawrence Freedman, The Evolution of Nuclear Strategy, Third Edition (New York: Palgrave-Macmillan, 2003), p. 461.

[24] Keith B. Payne, "Future of Deterrence: The Art of Defining How Much Is Enough," *Comparative Strategy*, No. 3 (July–August 2010), pp. 217–222, citation p. 219.

[25] Kenneth N. Waltz, "More May Be Better," Ch. 1 in Scott D. Sagan and Kenneth N. Waltz, *The Spread of Nuclear Weapons: A Debate* (New York: W.W. Norton, 1995), pp. 1–45. Counter-arguments by Scott Sagan and counter-counterarguments by Waltz and Sagan also appear in this volume.

systems theory for logical support, but Patrick M. Morgan warns that systemic trends after the Cold War might be less supportive of deterrence than many assume:

> The Cold War and nuclear weapons gave deterrence an undeserved good name. In prior periods the mixed utility of deterrence was much clearer in that it was often not very effective, sometimes counterproductive. That will be true now.[26]

One can argue the issues of credibility, reliability and morality of nuclear deterrence into a large bibliography without suffering the embarrassment of empirical disproof for favored arguments. However, conventional deterrence prior to the nuclear age, as well as after the arrival of nuclear weapons, does have a history and a pedigree. The track record for conventional deterrence is sufficiently dotted with failure that it must induce, on the part of historians, military planners and policy makers, considerable humility and respect for the indeterminacy of history. The "human factor" in time of war or crisis is capable of considerable "mystifying, misleading and misdirection," not only of opponents, but also of itself and its own side. As one academic expert on military strategy has opined, "Sometimes, I suspect that we survived the Cold War more by luck than judgment."[27]

Conclusion

New START reduced U.S. and Russian numbers of deployed strategic nuclear weapons and launchers below previously agreed limits, but not drastically so. The Obama administration's interest in New START was arguably more of a political than a military-technical exercise. New START displayed a practiced U.S.–Russian nuclear choreography, but within a post-Cold War political context. Additional accomplishments as between the United States and Russia were neither precluded nor guaranteed by New START. Obama's agenda for denuclearization was ambitious, but constrained by U.S. domestic politics, by NATO requirements for interalliance agreement, and by Russian willingness to reciprocate. New START was neither panacea nor placebo, but exactly what, remained to be defined.

New START required a trade-off as between U.S. and Russian interests and policy perspectives. Russia needs the reassurance provided by the START

---

[26] Patrick M. Morgan, *Deterrence Now* (Cambridge: Cambridge University Press, 2003), p. 263.

[27] Gray, "Gaining Compliance: The Theory of Deterrence and its Modern Application," p. 281.

regime, and the appearance of nuclear-strategic parity with the United States, in order to support its other political goals.[28] In turn, the United States wants to avoid a pointless nuclear arms race with Russia and to obtain Russian support for convergent, although not identical, political objectives as between Washington and Moscow.[29] These objectives include the continuing absence of major interstate war from Europe and further progress on European denuclearization; the management of nuclear nonproliferation, including the cases of Iran and North Korea; and, not least, cooperation against international crime and terrorism, including the possibility of terrorists armed with nuclear weapons and the containment of disruptive forces in Afghanistan.

---

[28] For perspective on Russian nuclear arms control and national security objectives, see: Stephen J. Blank, *Russia and Arms Control: Are There Opportunities for the Obama Administration?* (Carlisle, Pa.: Strategic Studies Institute, U.S. Army War College, March 2009), and Daniel Goure, "Russian Strategic Nuclear Forces and Arms Control: Déjà vu All Over Again," Ch. 5 in Blank and Weitz, eds., *The Russian Military Today and Tomorrow: Essays in Memory of Mary Fitzgerald* (Strategic Studies Institute, U.S. Army War College, July 2010), pp. 301–329.

[29] For informative perspective on this topic, see Alexei Arbatov, "Terms of Engagement: Weapons of Mass Destruction Proliferation and U.S.–Russian Relations," Ch. 5, pp. 139–168 and Stephen J. Blank, "Prospects for Russo-American Cooperation in Halting Nuclear Proliferation," Ch. 6, pp. 169–284, both in Stephen J. Blank, ed., *Prospects for U.S.–Russian Security Cooperation* (Carlisle, Pa.: U.S. Army War College, March 2009).

6

Nuclear threat and North Korea: Dangers and options

Introduction

In this chapter, the North Korean nuclear enigma is discussed as a problem in nuclear arms control and disarmament, and beyond that, as an especially challenging policy study in conflict termination. Discouraging of wishful thinking, the North Korean "problem" is admittedly dangerous, possibly manageable, but ultimately destructive to intellectual or policy complacency. The challenge presented by a nuclear North Korea causes policy and strategy concerns, and not only on account of the immediate and specific risks for the Korean peninsula or for the region of North Asia. The North Korean "problem" defies solution apart from consideration of the larger context within which nuclear danger is created in the present century. This larger context includes the debatable efficacy of deterrence, the degree of stress on the global nonproliferation regime, and the requirement for leadership in nuclear arms reductions that falls to the United States and Russia.

A North Korean challenge

U.S. President Barack Obama had no sooner proclaimed in favor of nuclear abolition as an American and global policy objective than he received a crude reminder of why abolition will be so difficult to accomplish.[1] In May and June, 2009 North Korea conducted its second acknowledged nuclear test and fired off a series of ballistic missiles in defiance of prior United Nations resolutions and U.S. warnings. North Korea's actions confounded expectations and agreements reached between Pyongyang and five negotiating partners during the George W. Bush administration (the United States, South Korea, Japan, Russia and China).[2]

The Obama administration reacted with undisguised disappointment to the North Korean testing and launching initiatives. U.S. officials considered interdicting North Korean air and sea shipments suspected of carrying weapons or nuclear technology. Also under examination was the option of putting North Korea back on the U.S. list of states that were officially designated as sponsors of terrorism. The Bush II administration had previously removed the DPRK from the list in 2008.

Cabinet level officials in the Obama administration expressed frustration and concern about the North Korean nuclear and missile tests, as well as about their implications for the nonproliferation regime in Asia and for the future of the six-party talks on North Korean denuclearization. U.S. Secretary of State Hillary Rodham Clinton warned that, if the North Korean demarches did not meet with a strong reaction from other states and international organizations, the risk of a larger arms race in northeast Asia was imminent.[3] Secretary of Defense Robert Gates was even more blunt than his counterpart at Foggy Bottom. With respect to prior agreements reached with North Korea on the issue of denuclearization, Gates stated: "I'm tired of buying the same horse twice."[4] Gates made this statement, probably not coincidentally, while visiting the U.S. ballistic missile defense (BMD) site at Fort Greely, Alaska, ostensibly deployed by the George W. Bush administration in response to threats from North Korea. Clinton's and Gates' frustration with the regime in Pyongyang

---

[1] Jennifer Loven, "Obama outlines sweeping goal of nuclear-free world," Associated Press, April 5, 2009, in *Johnson's Russia List* 2009 – #66, April 5–6, 2009, davidjohnson@starpower.net.

[2] Obama administration officials have expressed concerns about the possibility of a leadership crisis in North Korea. See "Clinton: U.S. fears N. Korea leadership crisis," Associated Press, February 19, 2009, http://www.msnbc.msn.com/id/29272217/html. Informed assessment of North Korean military capabilities appears in Andrew Scobell and John M. Sanford, *North Korea's Military Threat: Pyongyang's Conventional Forces, Weapons of Mass Destruction, and Ballistic Missiles* (Carlisle, Pa.: U.S. Army War College, Strategic Studies Institute, April 2007). Expert analysis of North Korean nuclear capabilities appears in Siegfried S. Hecker, "The risks of North Korea's nuclear restart," *Bulletin of the Atomic Scientists*, May 12, 2009, http://thebulletin.org/web-edition/features/the-risks-of-north-koreas-nuclear-restart.

[3] David E. Sanger, "U.S. Weighs Intercepting North Korean Shipments," New York Times, June 7, 2009, http:www.nytimes.com/2009/06/08/world/asia/08korea.html.

[4] Ibid.

reflected the attitude of the U.S. President himself. Aides said that President Obama had decided not to offer North Korea any new incentives to dismantle its nuclear complex at Yongbyon, the site of on-again, off-again dismantlement under agreements reached in the six-party negotiations. During his trip to Europe in June, 2009 Obama told reporters that he would not conduct negotiations with North Korea as in the past. "I don't think there should be an assumption that we will simply continue down a path in which North Korea is constantly destabilizing the region and we just react in the same ways," he said.[5]

Senior U.S. officials also said that the Obama administration would revisit and question the assumptions on which U.S. policy toward North Korea had been based for the past two presidential administrations of Bill Clinton and George W. Bush. Negotiations during these sixteen years assumed that the DPRK would eventually be willing to give up its nuclear weapons and dismantle its nuclear infrastructure in return for some combination of "carrots" supplied by the United States and other interlocutors: including economic assistance, assistance in developing a civilian nuclear power industry, and U.S. assurances against any attempted regime change in Pyongyang.

Instead, Obama advisors moved toward a different policy consensus about North Korean intentions and policy objectives. That new consensus privileged the assumption that North Korea desired to be recognized as a nuclear weapons state and that it had no intention of relinquishing or dismantling this capability. According to this reasoning, offering new incentives to North Korea to obtain fulfillment of agreements previously made and then suspended would send entirely the wrong signal. For example, with reference to prior agreements about the Yongbyon nuclear complex, where spent nuclear fuel had been reprocessed into bomb-grade plutonium, a key Obama advisor stated that, "Clinton bought it once, Bush bought it again, and we're not going to buy it a third time."[6]

This pessimism was all too prescient. In November, 2010 North Korea revealed the existence of a new facility at Yongbyon for uranium enrichment. This revelation confirmed American and other suspicions that the regime in Pyongyang was embarked upon a second path for building nuclear weapons. A U.S. nuclear expert who was among those permitted to see the new facility was impressed by how modern the technology was.[7] The expert, Dr. Siegfried S. Hecker, a former director of the Los Alamos

---

[5] Ibid.

[6] Ibid.

[7] David E. Sanger and Joseph Berger, "Arms Bid Seen in New N. Korea Plant," *New York Times*, November 21, 2010, http://www.nytimes.com/2010/11/22/us/22talk.html. See also: Siegfried S. Hecker, *A Return Trip to North Korea's Yongbyon Complex* (Stanford, Calif.: Center for International Security and Cooperation, Stanford University, November 20, 2010).

National Laboratory, reported that the plant was an "industrial-scale uranium enrichment facility with 2,000 centrifuges" and that its interior was "stunning" in its sophistication.[8] Dr. Hecker was told by North Korean officials that they had required only eighteen months for completion of their newly declared enrichment facility, although nuclear experts in South Korea doubted this estimate. In addition, some analysts also suspected that significant outside help had been made available to North Korea for this project. South Korean expert analysts suggested that North Korea might have smuggled banned nuclear equipment from Iran, through Afghanistan and perhaps Pakistan, and then through China and into North Korea.[9]

Within days of news reports about the uranium enrichment at Yongbyon, North Korea also fired artillery barrages at a South Korean island, Yeonpyeong, located in the Yellow Sea (or Western Sea) two miles from the Northern Limit Line (a disputed sea border that the North does not acknowledge). The shelling of the island, housing a garrison of about 1,000 South Korean marines and some 1,600 civilians, killed two South Korean soldiers and caused the South Korean military on November 22, 2010 to go into "crisis status."[10] North Korea blamed South Korea for firing test shots in the area during a South Korean annual military exercise, but the United States, Britain, Japan and South Korea dismissed this excuse and regarded the North Korean attack as a provocation. Russia and China called for restraint on the part of both Koreas, and China specifically emphasized the need to resume the six-party talks on North Korean denuclearization. Some experts in South Korea judged that the North Korean assertiveness was a sign of frustration in Pyongyang during a process of leadership transition. North Korean frustration, at the refusal of the Obama administration to remove sanctions imposed on the DPRK for its nuclear activities, might have prompted aggression against South Korea as a surrogate. As one South Korean expert on North Korea explained, faced with the inability to pressure Washington, North Korea had "taken South Korea hostage again."[11]

U.S. frustration with North Korea was understandable, but not original. Negotiating with this politically isolated, economically destitute and patrimonial

---

[8] Mark McDonald, "South Korea Digests News of North's Nuclear Site," *New York Times*, November 22, 2010, http://www.nytimes.com/2010/11/23/world/asia/23korea.html.

[9] Ibid. For a discussion of DPRK nuclear and missile programs in the context of regional military balances, see Anthony H. Cordesman, *The Korean Military Balance 2011: Comparative Korean Forces and the Forces of Key Neighboring States: Executive Summary* (Washington, D.C.: Center for Strategic and International Studies, revised May 6, 2011), esp. pp. 4–6 and 21–26.

[10] Mark McDonald, "Crisis Status in South Korea After North Shells Island," *New York Times*, November 23, 2010, http://www.nytimes.com/2010/11/24/world/asia/24korea.html.

[11] Ibid.

regime was difficult even in more propitious times for negotiators. However, the pessimism of the Obama administration may have been more tactical than existential. Officials may have decided that they had little to lose, given the experience of Bush and Clinton, by setting a tone of firmness and a policy of not offering preemptive concessions. The record since 1994 did not encourage optimism about the outcome of bilateral or multilateral negotiations with North Korea, but what was the alternative? [12]

As Michael Krepon has noted, the instruments for influence in the "first nuclear age" are essentially the same as in the "second nuclear age" following the end of the Cold War: deterrence; military strength; containment; diplomatic engagement; and arms control.[13] With regard to arms control (including what is now called cooperative threat reduction), some positive trends are apparent: the taboo against nuclear first use continues to hold;[14] nuclear weapons are decreasingly useful for major powers as their conventional military options improve; the global system to prevent nuclear and other weapons of mass destruction (WMD) proliferation has been resilient, despite some disappointments and setbacks; and, as yet, there have been no acts of nuclear terrorism, although experts and governments are justifiably fearful of the same.[15]

On the other hand, Krepon and other experts also acknowledge a number of negative, and potentially game changing, developments that might offset otherwise progressive trends in nonproliferation and demilitarization. These possibilities include: the first use of a nuclear weapon in warfare between states since Nagasaki; failure to stop and-or reverse the nuclear weapons programs in North Korea and Iran; inability to safeguard and secure nuclear materials and weapons against terrorist or other unauthorized use; an act of nuclear terrorism by al-Qaeda or other stateless groups; the spread of nuclear enrichment and reprocessing technologies and facilities to states that might

---

[12] Jonathan D. Pollack notes that U.S. negotiating with North Korea has alternated between coercive and incentive-based strategies without reconciling the two approaches. See Pollack, *North Korea's Nuclear Weapons Development: Implications for Future Policy*, Proliferation Papers, No. 33 (Paris: Security Studies Center, IFRI, Spring 2010), esp. p. 35, http://ifri.org/downloads/pp33-pollack.pdf.

[13] Michael Krepon, *Better Safe than Sorry: The Ironies of Living with the Bomb* (Stanford, Calif.: Stanford University Press, 2009), p. 174 and passim.

[14] On the concept of a nuclear taboo, see Nina Tannenwald, *The Nuclear Taboo: The United States and the Non-Use of Nuclear Weapons Since 1945* (Cambridge: Cambridge University Press, 2007), esp. pp. 327–360, and George H. Quester, *Nuclear First Strike: Consequences of a Broken Taboo* (Baltimore, Md.: Johns Hopkins University Press, 2006).

[15] Krepon, *Better Safe than Sorry*, pp. 135–137. On the risks and prevention of nuclear terrorism, see Graham Allison, *Nuclear Terrorism: The Ultimate Preventable Catastrophe* (New York: Times Books – Henry Holt and Co., 2004).

prefer to be a "screwdriver's turn" away from nuclear weapons status; and, finally, a possible collapse of the nuclear nonproliferation regime, as a result of these and other factors.[16]

Although the instruments of influence for dealing with nuclear danger, before and after the Cold War, are reasonable as outlined above, they are not necessarily predictable in their outcome. As applied to any specific case of decision making, including nonproliferation and nuclear arms control, the instruments of influence are subject to many uncertainties. These uncertainties include the ability of each side in any negotiating process to understand correctly the incentive structures and mind sets of the "other." The mind sets of national leaders are matters of considerable complexity, influenced by factors as obvious as professional education and training, or as subtle as deep psychology and perception. Therefore, the potential for erroneous judgments on the part of states in estimating the priorities and proclivities of other countries and their leaders is enormous. History is replete with misestimates and erroneous judgments that led to defeat in war or to flawed crisis management, including the "July crisis" of 1914 and the Cuban missile crisis of 1962.[17]

Even if the problem of influence is defined more narrowly, as having to do with deterrence per se, the uncertainties do not necessarily reduce themselves to manageable proportions. Deterrence is fundamentally subjective, having to do with the threat of force in order to obtain compliance with the demands of the threatener. Deterrent threats may be active or passive, explicit or implicit. United States and Soviet nuclear capabilities during the high Cold War constituted a threat in being: it was unnecessary for either side to make threatening noises about its capability for nuclear devastation. When, in contrast, Soviet leader Nikita Khrushchev turned to rocket diplomacy during the later 1950s and early 1960s, the result was his own embarrassment when the Kennedy administration leaked the facts of Soviet nuclear missile inferiority to the media. This embarrassment, in turn, might have provoked Khrushchev to further confrontation and even adventurism, in the form of nuclear missile deployments to Cuba.

---

[16] Krepon, *Better Safe than Sorry*, pp. 137–138. See also: Allison, *Nuclear Terrorism*, Ch. 7 and Joseph Cirincione, *Bomb Scare: The History and Future of Nuclear Weapons* (New York: Columbia University Press, 2007).

[17] For interesting case studies and pertinent theory, see: Paul Bracken, Ian Bremmer and David Gordon, eds., *Managing Strategic Surprise: Lessons from Risk Management and Risk Assessment* (New York: Eurasia Group for the National Intelligence Council, 2005); Eliot A. Cohen and John Gooch, *Military Misfortunes: The Anatomy of Failure in War* (New York: Random House – Vintage Books, 1991); and, James G. Blight and David A. Welch, *On the Brink: Americans and Soviets Reexamine the Cuban Missile Crisis* (New York: Hill and Wang, 1989).

If the Cold War Americans and Soviets, despite considerable practice, failed on occasion to understand one another's motives with respect to deterrence, it cannot be surprising that North Korea constitutes its own mini-version of Churchill's riddle, wrapped in a mystery, inside an enigma.[18] Pavel Podvig, Stanford University research physicist and leading authority on Russian nuclear forces, examines the issue of whether nuclear deterrence has helped to dissuade North Korea from becoming a nuclear weapons state.[19] He notes that deterrence theory, à la the Cold War, would anticipate that a powerful nuclear weapons state like the United States (or, presumably, Russia or China) could deter a North Korea or other threshold nuclear power from crossing the line. But in fact, it has not turned out that way. It may be that nuclear weapons "add nothing to existing nuclear weapons states' or the international community's abilities to prevent future North Koreas or to effectively confront them."[20]

How can this be? The answer is in at least two parts. First, the taboo against nuclear first use has an inertial force of more than sixty years behind it. As Nina Tannenwald has noted, this prescription against nuclear first use has survived the stress of the Cold War, and of the post-Cold War period, despite the fact that major powers in both periods held to declaratory policies allowing nuclear first use.[21] During the Cold War, the United States and NATO asserted the option of nuclear first use in order to deter a large scale Soviet conventional attack. After the Cold War, post-Soviet Russian military doctrine acknowledged the option of nuclear first use in some cases of conventional military operations on Russian state territory or otherwise posing a strategic threat to Russia. Thus both Cold War and post-Cold War experience suggest that the nuclear "taboo" is more than an operational norm or an extended habit. It acquired a moral or normative force that is to some extent independent of shifts in the nuclear balance of power or political alignments.

Another reason for the possible incapacity of nuclear deterrence to dissuade potential proliferators is the difficulty of proving that deterrence "worked" under exigent circumstances.[22] Proving that deterrence actually caused a

---

[18] For an assessment of North Korea's military potential, see Andrew Scobell and John M. Sanford. *North Korea's Military Threat: Pyongyang's Conventional Forces, Weapons of Mass Destruction, and Ballistic Missiles.* Carlisle, Pa.: U.S. Army War College, Strategic Studies Institute, April 2007.

[19] Pavel Podvig, "What if North Korea were the only nuclear weapon state?" *Bulletin of the Atomic Scientists*, May 27, 2009, http://thebulletin.org/web-edition/columnists/pavel-podvig/.

[20] Ibid.

[21] Tannenwald, *The Nuclear Taboo*, passim.

[22] On this and other issues related to deterrence theory and practice during the Cold War and after, see: Patrick M. Morgan, *Deterrence Now* (Cambridge: Cambridge University Press, 2003).

state to forego some decisions that it otherwise would have made requires the historical jiu-jitsu of "proving a negative." Historians and social scientists are rarely so fortunate as to have exemplary data or exhaustive documents to establish that leaders acted as they did because they were deterred by nuclear, or other, threats. Leaders' reasons for taking controversial decisions are rarely monocausal. And, in fact, deterrent and other threats are sometimes provocative of the very escalation that they seek to prevent. Khrushchev's threats against Presidents Eisenhower and Kennedy, to squeeze the allies out of West Berlin, only firmed the resolve of both leaders to reassert allied rights and to devise military plans that included, in extremis, the option for nuclear escalation.

Another uncertainty about deterrence, including nuclear deterrence, is the range of behaviors over which the manipulation of risk at the heart of deterrence can be said to be effective, or persuasive. This range of behavior, presumably subject to the influence of deterrence, might be described as the "domain" of deterrence, or as the "dependent variable" over which deterrent or other threats may be effective. However, this domain or set of dependent variables will vary considerably, based on factors such as the prevailing technology and its distribution among states, on the ranking of policy priorities by political leaders, on the preferential military doctrines of armed forces, and on other characteristics of the situation. To the matrix of dependent variables that might or might not be influenced by deterrent threats, we must add the *subjective expectations* about that matrix held by each side about the *other's* assessments. The reciprocal character of expectations is especially important in defining the conditions for success or failure in deterrence situations, as Thomas Schelling has explained in a number of his works.[23]

Another problematical aspect with respect to nuclear, or conventional, deterrence is the assumption that leaders from different states will share a transcultural means of interpretation and communication. Instead, leaders of state or nonstate actors may hold to different definitions of rational or sensible decision making than those taken for granted in United States or other Western cultures.[24] Terrorists motivated by apocalyptic or chiliastic definitions of their goals are one obvious example of persons whose calculus of deterrence

---

Colin S. Gray, *The Second Nuclear Age* (Boulder, Colo.: Lynne Rienner, 1999); Keith B. Payne, *Deterrence in the Second Nuclear Age* (Lexington, Ky.: University Press of Kentucky, 1996); Robert Jervis, *The Meaning of the Nuclear Revolution: Statecraft and the Prospect of Armageddon* (Ithaca, N.Y.: Cornell University Press, 1989); and Lawrence Freedman, *The Evolution of Nuclear Strategy* (New York: St. Martin's Press, 1981 and 1983).

[23] For example, see Thomas C. Schelling, *Arms and Influence* (New Haven, Ct.: Yale University Press, 1966).

[24] On this point, see especially Morgan, *Deterrence Now*, and Payne, *Deterrence in the Second Nuclear Age*, passim.

may be different from those of their targets. So widely acknowledged is the preceding point that it is an almost overstated cliché: terrorists and other unconventional warriors are "beyond deterrence." This assumption was used by the Bush II administration to justify not only preemptive attacks on terrorist leaders and cells, but also to permit preemptive or preventive strikes against certain regimes – including rogue states capable of launching WMD attacks or willing to aid terrorists in doing so.

Threats of preemptive attack or preventive war have the potential not only of deterring the intended recipients, but also of influencing their threat perceptions in unpredictable ways.[25] When the George W. Bush administration rhetorically conflated Iran, Iraq and North Korea into an "axis of evil," it appeared to put each of these states on notice that its regime survival could not be taken for granted. In addition, when the United States deposed the Saddam Hussein regime in Iraq in 2003, the governments in Iran and in North Korea were certainly more than interested bystanders. Regime change in Iraq, but not in North Korea, suggested that even a small nuclear capability could be deterring to the U.S. or other extraregional power with aspirations for regime change. Iran, having noticed the comparative fates of the Saddam Hussein and Kim Jong-il regimes, could have been encouraged to speed up its pursuit of a nuclear weapons capability or at least a "turnkey" complete nuclear fuel cycle.

Broadening the discussion

The preceding discussion might hold some important implications for those policy makers or analysts speculating about North Korea's intentions and about possibly effective, or potentially counterproductive, measures to discourage continuation on a path toward nuclear weapons status. Instead of assuming a singularity about North Korea and its intentions, suppose we generate a plurality of alternative North Koreas for analytic purposes?[26] For this purpose, a two-variable model of the "mind set" of North Korea's leadership might be a useful heuristic. The first variable is the degree to which the leadership is risk averse or risk acceptant. The second variable is the extent to which the leadership is politically secure or insecure within its own political milieu (Table 6.1).

---

[25] Contemporary and historical perspective on this issue are provided in Karl P. Mueller, et. al., *Striking First: Preemptive and Preventive Attack in U.S. National Security Policy* (Santa Monica, Calif.: RAND, 2006), and Colin S. Gray, *The Implications of Preemptive and Preventive War Doctrines: A Reconsideration* (Carlisle, Pa.: Strategic Studied Institute, U.S. Army War College, July 2007).

[26] Paul K. Davis, RAND, first suggested this idea to me in the context of a different study. He is not responsible for its application here.

Table 6.1 Alternative "North Koreas" leadership typologies

	Risk Acceptant	Risk Averse
Politically Secure	1	2
Politically Insecure	3	4

How would these leadership or regime typologies relate to a greater or lesser interest in nuclear weapons on the part of North Korea, or to its willingness to bargain away its nuclear capability for political or economic inducements? The most dangerous North Korea from the standpoint of regional stability and nonproliferation might be "North Korea 3" with its leadership genotype "politically insecure, risk acceptant." The most reasonable might be "North Korea 1" with politically secure and risk averse leaders.

United States and other negotiators with North Korea would certainly prefer to have a secure political leadership to deal with (in order to reach agreements that will last). The United States and some of North Korea's neighbors, including China, also fear a succession crisis in the DPRK that might erupt into civil war, with different military factions supporting different putative successors to Kim Jong-il. Civil strife in North Korea could send refugees pouring into China, South Korea or even Russia. In addition, under conditions of regime shakiness and intramilitary conflict, the secure custody of nukes could be jeopardized. The disintegration of the Kim family regime might be followed by the bizarre spectacle of the United States, China and Russia "cooperating" to restabilize North Korea, to disarm regional militants, and to re-secure or dismantle its nukes.[27]

The problem for the United States and other negotiators with North Korea, either on nuclear or other matters, is that it must prepare to deal with all four "North Koreas" and perhaps others not depicted here. The point is to avoid being fatalistically trapped in self defeating assumptions that contribute to any one of three dangerous fallacies: (1), wrongly assuming that North Korea is intransigent and hell bent on becoming a nuclear weapons state, when in fact the issue is still up for grabs in Pyongyang or is negotiable with a post-Kim Jong-il regime; (2), mistakenly assuming that diplomacy alone, supported by the proper mix of inducements and patience on the part of negotiators, will eventually bring North Korea to the conference table and to

---

[27] In this context, Andrew Scobell makes the important point that regime change in North Korea might take place without a total collapse of the state. For pertinent scenarios, see Scobell, *Projecting Pyongyang: The Future of North Korea's Kim Jong Il Regime* (Carlisle, Pa.: Strategic Studies Institute, U.S. Army War College, March 2008), esp. p. 25. See also: Young Whan Kihl and Hong Nack Kim, eds., *North Korea: The Politics of Regime Survival* (Armonk, N.Y.: M.E. Sharpe, 2006).

favorable agreement to denuclearize; or, (3), wrongly assuming that diplomacy has failed utterly, and further negotiating is useless – only threats of political and economic sanctions, or of military actions, will bring North Korea back to negotiations, and then to agreement.

Two approaches to "thinking out of the box" about U.S. and other negotiators' options with respect to North Korea might suggest themselves. First, negotiations with North Korea might adopt their own regional variation on diplomat and expert negotiator Dennis Ross's "bigger carrots, bigger sticks" approach planned for use by the Obama administration in the Middle East. Second, and peculiar to the Korean situation, the United States and others might broaden the definition of the political task. Instead of continuing talks on denuclearization, the United States and its partners in Asia might pursue the problem as one of war termination. The Korean War is, from a political standpoint, still without an end. Fighting on the Korean peninsula was terminated by an armistice that remains in place. A permanent political solution to the relationship between the two Koreas remains at large.

Including nuclear restraint on North Korea within a larger set of negotiations on termination of the Korean War might involve two phases. The first phase would include the United States, China and the two Koreas – the major combatants in Korea from 1950 to 1953. Final political terms of endearment between the two Koreas would be negotiated in this forum – including humanitarian and economic relief for starving North Koreans; political recognition of the DPRK by the United States with a pledge against any armed attack for the purpose of regime change; and an opening of negotiations for eventual reunification of the two Koreas. In a second phase, involving Japan and Russia as well as the United States, China and the two Koreas, the issue of denuclearization of the entire Korean peninsula could be addressed. Both the DPRK and South Korea could pledge to remain as non-nuclear weapons states (with or without unification). In return, the United States, China, Russia and Japan would agree to provide support for civilian nuclear power industries in both Koreas – with oversight by inspectors from their countries to guarantee that nuclear fuel cycles were not being used to produce bomb-grade material. As Leon V. Sigal has noted, deepening engagement "is the only way to encourage change in North Korea" and also "our only way to enhance our political leverage."[28] He adds:

> Pyongyang may be willing to trade away its plutonium, enrichment and missile programs brick by brick. Washington should be willing to offer it

---

[28] Leon V. Sigal, "Let's Make a Deal," *The American Interest Magazine*, January–February 2010, http://the-american-interest.com/article-bd.cfm?piece=767.

much more, in return for much more. That includes diplomatic recognition and an Obama–Kim summit meeting as the DPRK dismantles its fuel fabrication plant, reprocessing facility and reactor at Yongbyon and allows its plutonium declaration to be verified.[29]

Sigal also supports the principle of a new peace process ending in a peace treaty and peninsular denuclearization. In essence, he argues, as do we, for a nonlinear approach to negotiating with the DPRK, emphasizing a progression of interlocking commitments that create a sense of accomplishment and a positive feedback loop for further agreements. Admittedly, this optimistic scenario assumes that, at least to some extent, North Korea can be treated as a "normal" state or a rational actor. It also assumes the continuity, at least in the near term, of the Kim Jong-il regime, although uncertainties surround the health of the "Dear Leader." On the other hand, diplomatic-strategic prudence may dictate that the United States and other interlocutors deal with the devil they know, as opposed to what a post-Kim Jong-il regime might look like.[30] As of autumn, 2010 the process of regime succession had already begun with the increasingly public emergence and recognition of Kim Jong-un as the Dear Leader's favored successor.

The larger context: U.S.–Russian security cooperation and nonproliferation

The larger context, for solving or coping with the problem of North Korean nuclear proliferation, goes beyond an expansion of the context for denuclearization into one of war termination. The larger context includes the credibility and longevity of the entire global nonproliferation regime. This regime outperformed expectations during the Cold War and for the duration of the twentieth century. In the present century, it is already under stress from the spread of weapons technologies, from the aspirations of some states for nuclear weapons capabilities, and from nonstate actors seeking to acquire weapons of mass destruction. None of these problems is beyond solution or resolution, but they all require international political collaboration and diplomacy in support of cooperative security.

The creation of cooperative security space contributory to nonproliferation and disarmament begins with the United States and Russia. These two

---

[29] Ibid.

[30] Ibid. See also Scobell, *Projecting Pyongyang*, p. 25 and passim for possible outcomes.

states own more than 90 per cent of the world's nuclear weapons and have a responsibility to lead in the accomplishment of nuclear arms reductions, nonproliferation, and, according to some experts, nuclear abolition.[31] Since the end of the Cold War, Russia and the United States have made considerable reductions in their numbers of deployed nuclear weapons and launchers. They have also, under the Nunn–Lugar program in cooperative threat reduction, collaborated on the safety, security, accountability and dismantlement of former Soviet and Russian nuclear weapons. Their claim to nuclear leadership is uncontestable, but relations between the George W. Bush and Vladimir Putin administrations, especially during the second term of each president, stalled progress on a number of fronts. Frictions over U.S. plans for missile defenses in Eastern Europe, NATO enlargement, and Russia's disputes with Georgia, among other issues, slowed momentum toward further accomplishments in nonproliferation and nuclear arms reductions.

The Obama administration has indicated its intent to "reboot" U.S. relations with Russia, and Presidents Obama and Dmitry Medvedev have signed the New START treaty on strategic nuclear arms limitations. However, the United States and Russia cannot necessarily guarantee that their achievements in bilateral nuclear arms reductions will carry over into multilateral accomplishments of either of two types: (1), proportionate reductions in the sizes of their nuclear weapons inventories by other nuclear weapons states, and especially by other members of the P-5 cadre of permanent members of the UN Security Council (China, France and the United Kingdom); and, (2), additional support for strengthening of the nonproliferation regime on the part of existing nuclear weapons states, and in particular for a shared commitment to maintain the "Allison line" against no loose nukes, no new nascent nukes, and no new nuclear weapons states.[32]

Another issue that intrudes into the relationship between nuclear arms reductions ("vertical" disarmament or arms limitation) and nonproliferation ("horizontal" disarmament or arms limitation) is the question of missile defenses. Missile defense technologies have, on the evidence of history, notably underperformed against responsive threats even under conditions

---

[31] Arguments in favor of abolition appear in: Ivo Daalder and Jan Lodal, "The Logic of Zero," *Foreign Affairs*, No. 6 (November–December, 2008), pp. 80–95; George P. Shultz, William J. Perry, Henry A. Kissinger and Sam Nunn, "Toward a Nuclear-Free World," *Wall Street Journal*, January 15, 2008, p. A13; Jonathan Schell, *The Seventh Decade: The New Shape of Nuclear Danger*. (New York: Henry Holt and Co., 2007); and Krepon, *Better Safe than Sorry*, passim. Russian and U.S. responsibilities for leadership in nonproliferation are emphasized in Stephen J. Blank, *Russia and Arms Control: Are There Opportunities for the Obama Administration?* (Carlisle, Pa.: Strategic Studies Institute, U.S. Army War College, March 2009).

[32] Allison, *Nuclear Terrorism*, pp. 140–175 provides a more detailed explanation.

of test or simulation. The problem is not lack of ability or imagination on the part of weapons scientists and military planners. The difficulty for the defense is the asymmetry between the task of nuclear missile attackers and the challenge that faces antimissile defenders. The attacker need only succeed in penetrating the defense with a small number of missiles armed with nuclear warheads in order to create historically unprecedented, and politically unacceptable, disaster. On the other hand, the antimissile defenses must be nearly perfect in their ability to intercept attacking missiles and-or reentry vehicles in order to reduce damage to acceptable levels. (Another problem is that defenses against ballistic missiles do nothing to obviate strikes by cruise missiles or other forms of nuclear attack, but that issue is set aside for discussion purposes here.) Unless, or until, reliable systems for boost phase intercept are sustainably deployed, antimissile defenses will probably lag behind offensive countermeasures, including saturation attacks, deceptive simulation and antisimulation techniques for reentry vehicles, and hybrid cruise and ballistic missiles with adaptive trajectories.[33]

Although missile defenses that can protect national populations against large scale, nuclear missile attacks are beyond current or imminently foreseeable technology, a more feasible objective is the deployment of theater missile defenses (TMD) for the protection of U.S. allies or forward deployed forces. Theater missile defenses can provide some additional measure of deterrence by adding uncertainty to the calculations of attack planners. In addition, theater missile defenses, combined with a nuclear guarantee from an existing nuclear weapons state, can provide for dissuasion of prospective attackers both by means of denial and by the credible threat of retaliation. Among states prospectively in the gunsights of regional troublemakers, Israel and Japan benefit from the American "extended deterrence" umbrella as well as from the availability of theater missile defense technologies. In addition to

---

[33] The qualifier "sustainably deployed" is important. If, for example, antimissile interceptors were based on satellites deployed in space, then those satellites would have to be protected from antisatellite (ASAT) weapons by inherent self-defense capabilities or by assistance from "escort" satellites – as in the use of convoys and antisubmarine warfare (ASW) to protect surface shipping against submarine torpedo attacks. See also: Andrew M. Sessler, et. al. *Countermeasures: A Technical Evaluation of the Operational Effectiveness of the Planned US National Missile Defense System*. (Cambridge, Mass.: Union of Concerned Scientists, April 2000), and Dean Wilkening, *Ballistic-Missile Defence and Strategic Stability* (Oxford: Oxford University Press, 2000), pp. 25–27. Wilkening's analysis shows that airborne boost-phase intercept systems (ABI) may have the potential for high effectiveness against intercontinental and theater-range ballistic missiles launched by emerging ballistic missile states. The same systems would not necessarily pose a realistic threat to the Russian, Chinese or other major power deterrents. See Wilkening, *Ballistic-Missile Defense and Strategic Stability*, esp. pp. 60–0, and his illustration of ABI coverage of North Korea, p. 62.

those systems already deployed, technologies for theater missile defenses have been proposed that are based on available off-the-shelf technologies and would not be provocative of arms races. For example, UAVs (unpiloted airborne vehicles) armed with boost phase antimissile capabilities could loiter within or near dangerous regions, providing surveillance and prompt response to attack if necessary.[34]

U.S. hopes for avoiding a nuclear arms race in Asia rest, in part, on reassurance provided to Japan and to South Korea that denial and extended deterrent capabilities can substitute for a Japanese or South Korean nuclear force. Toward that end, U.S. and South Korean defense officials announced in December, 2010 the formation of an Extended Deterrence Policy Committee. Part of the Security Policy Initiative between the two governments, the extended deterrence committee would focus on countermeasures against North Korea's nuclear threat and other weapons of mass destruction, according to sources in the South Korean defense ministry.[35] South Korean officials also indicated that extended deterrence in this context meant that the United States could provide deterrence protection for South Korea by means of "tactical and strategic nuclear weapons, conventional strike and missile defense capabilities" for the defense of South Korea if attacked by North Korea.[36] The timing of the American and ROK announcement, about the creation of the extended deterrence committee, was doubtless not coincidental in view of recent North Korean provocations, including the DPRK's November 23, 2010 attack on the South Korean island of Yeonpyeong that killed four people.

Shifting sands of technology notwithstanding, missile defenses will not remove the majority of "ground zeros" for long range or other nuclear attacks from their hostage condition within the lifetime of the Obama or Medvedev administrations. Therefore, the cutting edge of a safer world, from the standpoint of mitigating nuclear danger, combines diplomacy with deterrence and nonproliferation – including supports for deterrence and nonproliferation such as cooperative threat reductions and arms control. Given that this is so, the existing nuclear weapons states must grapple with both vertical and

---

[34] UAVs could be used as launchers for airborne boost-phase intercept or strictly for improved target identification and early warning as part of a BMD architecture. For the latter role, see "UAV Concept Eyed for Longer-Term Missile Defense Applications," *Defense Daily*, April 18, 2003, reprinted in *Alaska Missile Defense Weekly*, 59th Edition, April 14–18, 2003, ronald.crowl@elmendorf.af.mil.

[35] "S. Korea, U.S. launch joint committee to deter N. Korea's nuclear threats," *Yonhap News Agency*, Seoul, December 13, 2010, http://app.yonhapnews.co.kr/YNA/Basic/Article/Print?YIBW_showEnArticlePrintView, downloaded December 15, 2010.

[36] Ibid.

horizontal disarmament in good time. This task is technically possible, but politically challenging with regard to the level of cooperation required among the powers.

Conclusions

United States and other negotiators, having been engaged in nuclear arms control talks with North Korea over the years, cannot be blamed for feelings of combat fatigue and diplomatic frustration. Yet no alternative to perseverance presents itself. Nuclear weapons in the hands of the Kim family regime, especially during a period of North Korean succession or during tensions between North and South Korea, are potential ignition points for multilateral troubles in Asia. The principal danger posed by North Korea might not be that of a nuclear first use or first strike: the result, if used against a nuclear power or an ally of a nuclear power, would invite the total destruction of the regime and its country.

Instead, North Korea would pose three other dangers: (1), attempted access denial with regard to military intervention in the region by outsiders, including the United States; (2), a Wal-Mart for spreading nuclear technologies, materials and know-how – including to terrorists; and, (3), a "failed state" following the disintegration of a weakened Kim regime and requiring the cooperation of the United States, China, Russia, and South Korea for security and stability operations, including the control of any loose nukes. None of these dangers is obviously amenable to solution by deterrence.

It is therefore inescapable that the United States and others who want to avoid a nuclear North Korea will have to engage diplomatically and relentlessly, with no guarantees of success. Meanwhile, construction of an alternate plan "B" in case of North Korean nuclear irreversibility, short of war, is time-urgent. As U.S. nuclear scientist Siegfried Hecker reported, after having visited North Korean nuclear facilities in November, 2010, waiting for Pyongyang to return to the six-party talks on terms acceptable to the United States might "exacerbate" the problem; military attack was "out of the question;" and, tightening sanctions further was "likewise a dead end." The only hope, according to Hecker, was further engagement.[37] As he noted in a subsequent article on this topic:

---

[37] Hecker, *A Return Trip to North Korea's Yongbyon Complex*, p. 8.

The fundamental and enduring goal must be the denuclearization of the Korean peninsula. However, since that will take time, the U.S. government must quickly press for what I call "the three no's – no more bombs, no better bombs, and no exports – in return for one yes: Washington's willingness to seriously address North Korea's fundamental insecurity along the lines of the joint communiqué.[38]

---

[38] Siegfried S. Hecker, "What I Found in North Korea," *Foreign Affairs*, December 9, 2010, http://www.foreignaffairs.com/print/66970, downloaded December 15, 2010. The joint communiqué was issued in October 2000 by the U.S. and North Korea, indicating that neither state had hostile intent toward the other and pointing toward a shared commitment to build a new relationship.

7

Nuclear "first use" and European peace: A risky bargain?

Introduction

The year 2010 marked the convening of NATO's historic Lisbon summit in November, attended by Russia and witnessing the adoption by NATO of a strategic concept that no longer defined Russia as a military threat. A corollary statement by the NATO–Russia Council also supported the idea that officially neither NATO nor Russia constituted a military threat to the other. In the NRC statement, NATO and Russian leaders pledged to work toward the achievement of "a true strategic and modernized partnership" based on reciprocal confidence, transparency and predictability and with the object of creating a "common space of peace, security and stability in the Euro-Atlantic area."[1] At the same summit meetings, NATO heads of state and government agreed to develop and deploy a NATO European-wide ballistic missile defense system to protect populations and territory. NATO also indicated its intent to invite Russia to take part in planning, and eventually implementing, the European BMD system, which would also be correlated with the U.S. European-based Phased Adaptive Approach to regional missile

---

[1] NATO, "NATO–Russia set on path towards strategic partnership," November 20, 2010, http://www.nato.int/cps/en/natolive/news_68876.htm.

defenses previously announced by the Barack Obama administration.[2] Meanwhile, across the Atlantic, the U.S. Congress was faced with a decision about whether to consent to ratification of the New START agreement for U.S.–Russian strategic nuclear arms reductions signed by Presidents Barack Obama and Dmitri Medvedev in April, 2010.[3]

NATO's strategic concept of 2010 and New START left for future discussion the thorny question of U.S. nuclear weapons deployed in five NATO countries and Russia's other than strategic nuclear weapons located within its European theater of military operations.[4] Variously referred to as substrategic, tactical or battlefield weapons, NATO's other than strategic weapons involve both arms control and nonproliferation issues for the United States and NATO, for Russia, and for other states that might have deployed weapons of shorter than intercontinental range on various delivery vehicles. Among the arms control issues associated with substrategic weapons (our preferred nomenclature), the problem of nuclear "first use" takes pride of place. Nuclear first use is the emblematic failure for substrategic nuclear deterrence, just as nuclear first strike is the signal failure for strategic nuclear deterrence.

However, the two types of deterrence and arms control failure, first use and first strike, are related in theory and in practice. Among other relationships, a decision for nuclear first use can initiate a sequence of events that result in a larger, and more destructive, nuclear first strike. Declaratory policies of nuclear first use or first strike, even if intended for deterrence or reassurance, carry prospective costs and risks. The risks of nuclear first use policies on the part of NATO or Russia might increase if the spread of nuclear weapons, especially in Asia, is not contained within present boundaries. In addition, the unfortunate possibility of ambiguous lines between nuclear first use and first strike, and equally indistinct markers between preemption and preventive war, has the potential to turn one state's deterrent into another's provocation.

---

[2] NATO, "Allied leaders agree on NATO Missile Defense system," November 20, 2010, http://www. nato.int/cps/en/SID-63044043-640CF22E/natolive/news_68439.htm. Pertinent expert commentary on the Lisbon summit with specific reference to missile defense appears in: Dmitry Trenin, "Turning a Happy Hour Into a Happy Alliance," *Moscow Times*, November 22, 2010, in *Johnson's Russia List* 2010 – #218, November 22, 2010; and James Sherr, "NATO and Russia: 'Refresh' but no Transformation," *Chatham House*, www.chathamhouse.org.uk, November 10, 2010, in *Johnson's Russia List* 2010 – #218, November 22, 2010.

[3] "Pressure Builds for New START Approval," www.globalsecuritynewswire.org, November 22, 2010, in *Johnson's Russia List* 2010 – #219, November 23, 2010.

[4] For an assessment of the pros and cons of maintaining, consolidating or downsizing NATO's sub-strategic nuclear weapons after the Lisbon summit, see Paul Schulte, *Is NATO's Nuclear Deterrence Policy a Relic of the Cold War?* (Washington, D.C.: Carnegie Endowment for International Peace, Policy Outlook, November 17, 2010).

SubStrategic, but not subImportant

Russia's nuclear weapons are, as prior discussion has established, its principal claims to military great power status unless, and until, its conventional forces are substantially approved. And quite understandably, given U.S. President Obama's call for a global nuclear zero and the emphasis placed by his administration on the New START arms reduction talks, media and public attention have focused on the limitation and ultimate fate of "strategic" nuclear weapons as opposed to others. Customary usage by arms control experts and military professionals refers to "strategic" nuclear weapons as those carried by missiles or bombers with intercontinental or transoceanic range – that is, more than 5,500 kilometers (the outer limit for intermediate range weapons, one step below strategic). Some have also attempted to distinguish strategic nuclear weapons from others on the basis of their respective yields, but this practice is less reliable or consensual among experts. A working compromise will be offered here. Strategic nuclear weapons are those operationally deployed or stored for possible deployment on launchers capable of ranges greater than 5,500 kilometers, or those deployed on United States, Russian or other states' intercontinental delivery systems, according to the established protocols of agreed arms control treaties or other verifiable sources.

For future reference here, nuclear weapons other than strategic ones will collectively be referred to as substrategic. We will stipulate that this category subsumes nukes referred to in various media as "battlefield," "tactical," "theater," and so forth – weapons intended for deployment on delivery systems of less than strategic, that is, intercontinental range.[5] During the Cold War, the possibility of mutual nuclear annihilation between the United States and Russia focused most arms control and military discussion on strategic nuclear weapons and delivery systems. Notable exceptions included the Intermediate Nuclear Forces treaty between the United States and the Soviet Union in 1987 (eliminating ground-launched ballistic and cruise missiles of 500–5,500 km ranges), and the unilateral-reciprocal Presidential Nuclear Initiatives (PNI) of 1991–1992 between President George H.W. Bush and his Soviet counterpart Mikhail Gorbachev and Russian counterpart Boris Yeltsin. Although the PNI agreements were not a formal treaty and included no provisions for verification, they resulted in the destruction of thousands of

---

[5] A very intelligent discussion of this issue appears in Richard Weitz, "Strategy and Doctrine in Russian Security Policy," paper presented at conference on "Strategy and Doctrine in Russian Security Policy," Ft. McNair, National Defense University, Washington, D.C., June 28, 2010, p. 19 and passim.

Russian and American substrategic nuclear weapons and the removal into storage of many others.[6]

One important difference between strategic and substrategic nuclear weapons, based on Cold War and subsequent experience, is that many substrategic weapons are intended for forward deployment with military forces in the field – such as divisions, tactical air wings, and fleets. They are embedded with the troops, so to speak, at the tip of the spear where conflict is likely to start. Strategic nuclear weapons are cosseted in specialized units, apart from "regular" forces, and operating according to their own set of training regimens and protocols. In addition, strategic weapons are likely to be weapons of last resort, after other options have been exhausted – including other nuclear ones.

Although all American, allied NATO and Soviet Cold War nuclear weapons, including substrategic weapons, were supposedly under strict centralized control of their respective national governments, substrategic operational and tactical weapons were deliberately deployed with publicly ambiguous policy guidance as to exactly when, and under what conditions, they might be activated. This aspect of substrategic weapons deployment might have been intentional, for the purpose of reinforcing deterrence, but it opened the door to misperception during a crisis. For example, a NATO nuclear command post exercise in early November, 1983, called Able Archer, may have unexpectedly played into Soviet fears of a U.S. plan for a nuclear first strike, a misperception of Reagan administration thinking that was stoked by elements in Soviet intelligence and by recent political controversies over thr NATO "572" missile deployments, President Reagan's "Star Wars" speech in March, 1983 and the Soviet shootdown of KAL 007 in September of the same year.[7]

The end of the Cold War, the demise of the Soviet Union, and the advent of the twenty-first century have not witnessed the elimination of United States, NATO or Russian substrategic nuclear weapons. To the contrary, the United States still deploys hundreds of substrategic nukes on the state territories of NATO allies, and Russia's large arsenal of substrategic nuclear weapons assuredly includes many deployed in its western military districts, intended for possible use in case of an attack by NATO. NATO's Group of Experts, tasked to prepare guidelines for the alliance's new Strategic Concept in 2010, reiterated the need for a nuclear component of NATO's deterrent, including U.S.

---

[6] Ibid., p. 20.

[7] Christopher Andrew and Oleg Gordievsky, *KGB: The Inside Story of Its Foreign Operations from Lenin to Gorbachev* (New York: Harper Perennial, 1991), pp. 590–600.

substrategic nuclear weapons deployed abroad.[8] Implicit in these American, allied NATO and Russian decisions with respect to substrategic nuclear weapons was another decision: to accept an unknown, but not insignificant, degree of risk of nuclear first use, in all likelihood developing on the heels of an outbreak of conventional war.

The second decision by the United States, NATO and Russia, to accept an unknown, but not insignificant, degree of risk of nuclear first use, became a discussion item among Russians in the run-up to the publication of Russia's revised Military Doctrine of 2010. In a speech at Russia's Academy of Military Sciences on January 19, 2008, General Yuri Baluyevksy, then Chief of the General Staff of the Russian armed forces, noted that Russia would use its military power to uphold its interests in a variety of situations. He emphasized that, if necessary, Russia would strike preemptively, not excluding the possible use of nuclear weapons in a first strike. According to Baluyevsky:

> We are not going to attack anyone, but we want all our partners to realize that Russia will use armed force to defend its own and its allies' sovereignty and territorial integrity. It may resort to a pre-emptive nuclear strike in cases specified by its doctrine.[9]

Experts immediately cautioned that Baluyevsky was restating the "traditional" position of Russia since the end of the Cold War, and also consistent with the 2000 military doctrine of the Russian Federation. In contrast to the Cold War declaratory policy of the Soviet Union, Russia's military doctrine included the option of nuclear first use or first strike in a conventional war involving attacks on Russian state territory or otherwise threatening to Russia's vital interests.

On the other hand, it was possible to interpret Baluyevsky's statement as a more assertive affirmation of the right of nuclear first use than hitherto for Russia's military command. The question remained open with respect to the particular circumstances of an attack and how Russia would define its "interests" and "sovereignty" as having been affected. Former Russian Defense Minister Sergei Ivanov reportedly considered as quite defensible the carrying out of presumably preemptive or preventive nuclear strikes against

---

[8] *NATO 2020: Assured Security; Dynamic Engagement, Analysis and Recommendations of the Group of Experts on a New Strategic Concept for NATO* (Brussels: North Atlantic Treaty Organization, May 17, 2010), pp. 43–44.

[9] Baluyevsky, quoted in Andrei Kislyakov, "Russian army prepares for nuclear onslaught," *RIA Novosti*, January 29, 2008, in *Johnson's Russia List* 2008 – #20, January 29, 2008, davidjohnson@starpower.net.

terrorists. Other high ranking Ministry of Defense officials have also discussed this option.[10]

In a more recent commentary on this issue, Russian Security Council Secretary Nikolai Patrushev, anticipating the release of the third edition of Russia's post-Soviet military doctrine, stated in November, 2009 that Russia views its nuclear arsenal first and foremost as a deterrent. However, he added, the option of a retaliatory or even preemptive nuclear strike was available in a situation defined as critical by Russia's leadership.[11] Patrushev further explained:

> The possibility of using nuclear weapons depends on the situation and intentions of the potential adversary. In critical situations for the national security a nuclear strike at the aggressor, including preventive strike, is not ruled out.[12]

Patrushev also made it clear that a potential adversary of Russia should understand the "futility of unleashing aggression with the use of nuclear and conventional means of destruction."[13] The "and" in this case might well be construed as an "or" – either nuclear or conventional aggression might prompt a nuclear response from Russia. Doubtless, some of this was boiler plate reiteration of huffing and puffing designed to compensate for the weakness of Russia's conventional forces, compared to NATO's. That weakness was on display in early August, 2008 during Russia's conflict with Georgia over South Ossetia. The clash resulted in an apparently one sided Russian "win" – albeit with disconcerting accounts of Russia's performance in the field. With, or without, Georgia on its mind, Russia's leadership was certainly mindful of the slow pace of its military reform, toward a more professional and rapidly deployable armed force suited to coping with unconventional conflicts, terrorism and other challenges of the present century.

---

[10] Vladimir Ivanov, "Comparison of Russian, U.S., and NATO Policies Toward Preemptive Nuclear Strikes," *Nezavisimoye Voyennoye Obozreniye*, February 4, 2008, in *Johnson's Russia List*, 2008 – #25, February 5, 2008, davidjohnson@starpower.net.

[11] "New Russian Doctrine Allows Preventive Nuclear Strike," *Itar-Tass*, Moscow, November 20, 2009, in *Johnson's Russia List* 2009 – #213, November 20, 2009, davidjohnson@starpower.net. An earlier press report on this topic appeared in David Nowak, "Report: Russia to allow preemptive nukes," Associated Press, October 14, 2009, in *Johnson's Russia List* 2009 – #190, October 15, 2009, davidjohnson@starpower.net. Analysis is presented in "Russia's Message on Reshaping Its Nuclear Doctrine," Stratfor.com, October 15, 2009, in *Johnson's Russia List* 2009 – #190, October 15, 2009, davidjohnson@starpower.net.

[12] Ibid.

[13] Ibid.

When, after numerous delays and false starts, Russia's new military doctrine was signed and published on February 10, 2010, it offered a more nuanced and reassuring perspective on the role of nuclear weapons in Russian military strategy and state policy. The 2010 doctrine emphasized that nuclear weapons existed primarily for the purpose of strategic deterrence and the avoidance of nuclear war. The doctrine acknowledged that, in the case of regional or larger conventional wars, the possession of nuclear weapons "may lead to such a military conflict developing into a nuclear military conflict" and also noted:

The Russian Federation reserves the right to utilize nuclear weapons in response to the utilization of nuclear and other types of weapons of mass destruction against it and [or] its allies, and also in the event of aggression against the Russian Federation involving the use of conventional weapons when the very existence of the state is under threat.[14]

The 2010 military doctrine stopped short of references to the use of nuclear weapons in local wars and seemed to place less emphasis on nuclear weapons altogether than pessimists had feared.[15] Nevertheless, and consistent with earlier formulations of post-Soviet Russian military doctrine, the option of nuclear first use was not precluded.[16] Also consistent with prior Russian military doctrine in 2000 was the assumed potential for substrategic nuclear weapons to be employed in response to some kinds of conventional attacks, for the purpose of strategic de-escalation – that is, for limited use to inflict "calibrated" damage with the objective of obtaining escalation

---

[14] Text, "The Military Doctrine of the Russian Federation," www.Kremlin.ru, February 5, 2010, in *Johnson's Russia List* 2010 – #35, February 19, 2010, davidjohnson@starpower.net. According to Dale R. Herspring, the assumption that Russia's nuclear deterrent can compensate for weaknesses in its conventional forces, come what may, is questionable, in view of the dubious quality of one or more legs of Russia's strategic nuclear triad. Herspring, "Russian Nuclear and Conventional Weapons: The Broken Relationship," paper presented at conference on "Strategy and Doctrine in Russian Security Policy," Ft. McNair, National Defense University, Washington, D.C., June 28, 2010.

[15] This and other aspects of the 2010 Military Doctrine are well analyzed in Keir Giles, "The Military Doctrine of the Russian Federation 2010," *NATO Research Review*, (Rome: NATO Defense College, February 2010), esp. p. 2 and pp. 5–6.

[16] For expert analysis of the 2010 Russian military doctrine, see: Roger N. McDermott, "Russia's Conventional Armed Forces, Reform and Nuclear Posture to 2020," paper presented at conference on "Strategy and Doctrine in Russian Security Policy," Ft. McNair, National Defense University, Washington, D.C., June 28, 2010; Marcel de Haas, "Russia's New Military Doctrine: A Compromise Document," in *Russian Analytical Digest*, No. 78, May 4, 2010, pp. 2–4, www.res.ethz.ch; and Nikolai Sokov, "The New, 2010 Russian Military Doctrine: The Nuclear Angle," Center for Nonproliferation Studies, Monterey Institute of International Studies, February 5, 2010, http://cns.miis.edu/stories/100205_russian_nuclear_doctrine.htm.

dominance and a favorable military outcome, while avoiding expansion of the conflict into a global war or a nuclear war between the United States and-or NATO and Russia.[17] The expectation of strategic de-escalation by means of limited nuclear use is interesting when paired with the doctrine's apparent obliviousness about Russian military reform. Expert analyst Keir Giles has noted:

> Despite the fact that the "new look" of the Russian Armed Forces is now closer to reality than ever before following the first real overhaul in post-Soviet history, reading the 2010 Military Doctrine you could be forgiven for assuming that nothing has changed at all.[18]

However, Russia was not alone in revisiting the spectrum of rationales for nuclear first use during the endgame of the George W. Bush and Vladimir Putin presidencies. Similar discussions about nuclear preemptive or preventive attacks were taking place in Western circles.[19] In a report prepared by five prominent former U.S. and allied NATO generals calling for "root and branch" reform of the alliance, the authors contend that NATO must be ready to resort to a preemptive nuclear attack to halt the "imminent" spread of nuclear and other weapons of mass destruction.[20] The authors, including retired General John Shalikashvili, former Chairman of the U.S. Joint Chiefs of Staff and former Supreme Allied Commander, Europe (SACEUR), and counterparts from Britain, France, Germany and the Netherlands, contended that a "first strike" nuclear option remained an "indispensable instrument" since there was "simply no realistic prospect of a nuclear-free world."[21] In a possibly oxymoronic or fatalistic construction, with regard to future NATO options, the authors noted that, "The first use of nuclear weapons must remain in the

---

[17] Nikolai Sokov, "Nuclear Weapons in Russian National Security Strategy," paper presented at conference on "Strategy and Doctrine in Russian Security Policy," Ft. McNair, National Defense University, Washington, D.C., June 28, 2010, pp. 13–17 on the concept of limited nuclear use with calibrated damage for the purpose of de-escalation.

[18] Giles, "The Military Doctrine of the Russian Federation 2010," p. 3.

[19] Strictly speaking, preemption is the decision to strike first on the basis of actionable intelligence that an enemy attack is imminent and unavoidable. Preemption is thus an act of striking first in the last resort. In contrast to preemption, preventive attacks are undertaken to forestall a possible, but not necessarily inevitable, future enemy attack or to reduce the power of a foreseeable future adversary. For an expansion, see Karl P. Mueller, Jasen J. Castillo, Forrest E. Morgan, Negeen Pegahi, and Brian Rosen, *Striking First: Preemptive and Preventive Attack in U.S. National Security Policy* (Santa Monica, Calif.: RAND, 2006), passim.

[20] Ian Traynor, "Pre-emptive nuclear strike a key option, NATO told," *Guardian Unlimited*, January 22, 2008, http://www.guardian.co.uk/nato/story/0,,2244782,00.html.

[21] Ibid.

quiver of escalation as the ultimate instrument to prevent the use of weapons of mass destruction."[22]

As in the case of Baluyevsky's and Patrushev's statements about Russian doctrine, cited above, the NATO generals' manifesto about nuclear first use can be interpreted in either of two ways: as a restatement, perhaps with brio, of existing doctrine; or, to the contrary, as a slight movement of the pendulum of usable military options further away from the "nuclear taboo" and toward an explicit preference for nuclear preemption or prevention under certain conditions.[23] The implication that either NATO or Russia might authorize the first use of nuclear weapons against nonstate actors who were planning attacks with weapons of mass destruction (WMD), and-or against states harboring such terrorists, was not unknown in military planning studies. But the public advertisement for such drastic military options seemed, at least for a time, to reach a higher decibel of recognition outside of professional military circles.

Russian, U.S. and NATO declaratory and operational policies with respect to nuclear first use are of interest not only to their respective internal audiences. Other state actors, including those with nuclear weapons and long range delivery systems, will take note. For example, China's official policy with respect to the use of nuclear weapons is one of "no first use." On the other hand, new doctrine for the use of missiles in warfare notes that a strategy of "active defense" can include sudden "first strikes" in campaigns or battles as well as "counterattacks in self defense" into enemy territory.[24] In addition, a vigorous debate has appeared among Chinese military and civilians about the viability of China's "no first use" policy, partly in the context of U.S. conventional military capabilities for a long range, precision strike against Chinese nuclear forces. According to one American expert on the Chinese military:

> They [People's Liberation Army military thinkers] fear that a conventional attack on China's strategic missile forces could render China vulnerable and leave it without a deterrent. This has led to a debate in China among civilian strategic thinkers and military leaders on the viability of the announced "no-first-use" policy on nuclear weapons. Some strategists advocate departing

---

[22] Ibid.

[23] The concept of a nuclear taboo and scenarios under which it might be violated are explored in George H. Quester, *Nuclear First Strike: Consequences of a Broken Taboo* (Baltimore, Md.: Johns Hopkins University Press, 2006).

[24] Larry M. Wortzel, *China's Nuclear Forces: Operations, Training, Doctrine, Command, Control, and Campaign Planning* (Carlisle, Pa.: U.S. Army War College, Strategic Studies Institute, May 2007), p. 9.

from the "no-first-use" policy and responding to conventional attacks on strategic forces with nuclear missiles.[25]

A further concern for U.S. military observers is the apparent mixing of nuclear, nuclear-capable and conventionally armed missiles within the same operational and tactical units. As Dr. Larry Wortzel, a U.S. Army War College expert on the Chinese military, has noted, the decision "to put nuclear and conventional warheads on the same classes of ballistic missiles and collocate them near each other in firing units of the Second Artillery Corps also increases the risk of accidental nuclear conflict."[26] Related to this concern about accidental or inadvertent nuclear war or escalation are the doctrinal emphases in PLA and Second Artillery thinking on the massing of decisive missile fires with surprise in a theater war; ambiguity about the kinds of warheads used in ballistic missile attacks on naval battle groups; and, third, increasing Chinese interest in the military uses of space and in capabilities for attacking U.S. C4ISR systems supporting warning, command-control and missile defense.[27]

Dimensions of the problem

The interest in Russia and NATO in the possibility of preemption, and in making more explicit the existence of preemption against terrorists or other nonstate actors, is quite understandable. In the aftermath of 9/11 and other high profile terrorist attacks in the United States and Europe, the "war on terror" has carried NATO military operations into Afghanistan and realigned U.S. military thinking and planning along the lines of asymmetrical warfare. Russia, also

---

[25] Ibid., pp. viii–ix.

[26] Ibid., p. 31. China's Second Artillery Corps is, in its nuclear aspect, somewhat equivalent to former Soviet Strategic Rocket Forces or the current Russian Strategic Missile Forces. The Chinese Second Artillery Corps is technically a PLA "branch," compared to the Army, Navy and Air Force, which are "services." The conventional missile forces are the most mature and dynamic of the Second Artillery's forces and are more numerous than its nuclear missiles. See Evan S. Medeiros, "Minding the Gap: Assessing the Trajectory of the PLA's Second Artillery," Ch. 4 in Roy Kamphausen and Andrew Scobell, eds., *Right-Sizing the People's Liberation Army: Exploring the Contours of China's Military* (Carlisle, Pa.: U.S. Army War College, Strategic Studies Institute, September 2007), pp. 143–189, esp. endnote 2, p. 182.

[27] Wortzel, *China's Nuclear Forces*, pp. 31–33 and passim. See also: Medeiros, "Minding the Gap," pp. 166–167, and Wortzel, "PLA Command, Control, and Targeting Architectures: Theory, Doctrine, and Warfighting Applications," Ch. 5 in Kamphausen and Scobell, eds., *Right-Sizing the People's Liberation Army*, pp. 191–234. Also valuable and unique is Lonnie D. Henley, "War Control: Chinese Concepts of Escalation Management," Ch. 5 in Andrew Scobell and Larry M. Wortzel, eds., *Shaping China's Security Environment: The Role of the People's Liberation Army* (Carlisle, Pa.: U.S. Army War College, Strategic Studies Institute, October 2006), pp. 81–103.

victimized by costly terrorist attacks since 9/11 and fighting against terrorists and insurgents in Chechnya, is as concerned as the United States and allied NATO countries about the possible use by terrorists of WMD. Both NATO and Russian leaders recognize that nuclear weapons in the hands of terrorists create an unacceptable risk of a catastrophic attack against their societies.[28]

Acknowledgement of the peril created by terrorists with nuclear weapons, or other weapons of mass destruction, does not necessarily lead to a conclusion in favor of *nuclear* preemption against such targets. There are several points to be considered in this regard. First, the United States now holds the high cards with respect to long range, conventional precision strike capabilities, supported by mastery of the information and electronics spectra and of technologies for C4ISR (command, control, communications, computers, intelligence, surveillance and reconnaissance). Given accurate intelligence and targeting information, the United States and therefore NATO can strike across continents or oceans and against virtually any target on the earth's surface with near impunity and unprecedented accuracy.

Second, nuclear weapons have collateral damage that may be unacceptable to the user. The first use of nuclear weapons in anger since Nagasaki will bring international inquiries, and possibly recrimination, for the perpetrator. Even tactical or "mini" nuclear weapons will cause civilian casualties in unknown numbers. And if, in the aftermath of a nuclear preemption for the sake of counterterrorism, the target is misidentified or the intelligence is at all flawed, the damage to the credibility of the attacker, in political and in moral terms, will be inestimable. For example, a preemptive *nuclear* attack on the pharmaceutical plant in Sudan in 1998, allegedly in cahoots with al-Qaeda and engaged in making or storing biological weapons, would have been worse than an embarrassment, given the ultimately ambiguous and widely disputed intelligence in support of that strike in 1998.

Some contend that more precisely delivered nuclear weapons with reduced yields are ideal for "bunker busting" against terrorist or rogue state actor storage facilities for WMD. Nuclear weapons would have the advantage of burning up the residue of any chemical or biological weapons stored at the suspect site. However, the collateral damage to surrounding communities and facilities might still be extensive, and the distribution of radioactivity across the region would be subject to a number of uncertainties, including weather and seasonal variations in climate. The collateral damage from reduced yield nuclear weapons might well exceed the expectations of optimists and, in the process, also bring into question American or NATO motives and ethics.

---

[28] Graham Allison, *Nuclear Terrorism: The Ultimate Preventable Catastrophe* (New York: Time Books – Henry Holt and Co., 2004).

It might be objected here that Russia, lacking the conventional military capabilities of the United States and NATO, has a stronger case for nuclear preemption against anticipated WMD attacks by terrorists. However, in carrying out a nuclear preemption, Russia faces some of the same decision making trade-offs as NATO does – and possibly others.[29] If Russia were to fire the first nuclear weapon since 1945 against terrorists, Russia's neighbors and trading partners would hold their breath. They would worry whether this was a sign of Russian willingness to repeat the exercise under conditions of similar, or lesser, provocation. The United States and NATO would be discussing whether to increase their own preparedness for nuclear war and the adequacy of their current forces for nuclear deterrence. Russia's economic relations with Western Europe could be destabilized and the Kremlin's program for building an entirely more prosperous economy based on energy sales might be disrupted. In addition, Russia's inclusion among the G-8 powers as an interlocutor rests not only on its raw economic or military power, but also on its perceived legitimacy and commitment to world order, that is, soft power.

Finally, there are the important particulars of any nuclear first use: against whom, and where. If Russia were to employ tactical or smaller nuclear weapons against terrorists on its own state territory, and if evidence proved that a terrorist WMD attack was indeed imminent, then the world would take notice, but the matter would be widely regarded as a justified self defense of Russia's homeland. More complicated would be the situation if Russia struck preemptively with nuclear weapons against alleged terrorists in its "near abroad," especially in states that are in contention with Russia over various issues or being considered for membership in NATO. A preemptive nuclear attack outside of Russia's own territory against terrorists, however threatening they are perceived to be, raises issues of violation of state sovereignty and sets the dangerous precedent that others can cross state boundaries in nuclear preemption of suspect terrorists.

Neither NATO nor Russia faces easy issues, therefore, in deciding whether and when to use nuclear preemption, whether first use or first strike. Indeed, the distinction between first use and first strike is itself a problematical aspect of the case for nuclear preemption. This conceptual problem exists alongside another: the relationship between preemption and preventive war. Discussion of these two distinctions and their implications follows in the next section.

---

[29] See Jacob W. Kipp, "Russia's Tactical Nuclear Weapons and Eurasian Security," *Jamestown Foundation Eurasia Defense Monitor*, March 5, 2010, in *Johnson's Russia List* 2010 – #46, March 8, 2010, davidjohnson@starpower.net, for expert analysis.

Preemption, prevention, first strike, first use

Preemption and prevention

The distinction between preemptive and preventive attacks lies in the attribution of motive, by the defender against the attacker; in the reliability of the intelligence, relative to the plans of the attacker; and in the time available for making decisions about whether an attack is in progress or being considered in good time. If a defender has actionable intelligence that an attack has already been set in motion or is imminent, then preemption is a means of avoiding the worst effects of being surprised. Of course, people can quibble about what "actionable intelligence" means, but for the present discussion it means verifiable information from human or technical (or both) sources that an attack is in progress or is about to be launched. For example, the U.S. nuclear attack warning system during the Cold War required confirmation by "dual phenomenology" (satellites and ground stations) before authoritative interpretation of an attack in progress was validated.

In addition to the reliability of the defender's intelligence about the attacker's capabilities and plans, the matter of time is also important in the justification for preemption. Preemptive attacks assume that the option of forestalling the attack by diplomacy or deterrence no longer exists. The attacker has taken an irrevocable political decision for war. The defender's options are to await the first blow, or, alternatively, to act first to minimize damage or, if possible, to preemptively destroy the enemy's strike capabilities. The time pressure for making these decisions creates a compression factor that can destabilize rational or even sensible decision making. Even when nuclear weapons are not involved, crisis management often brings out the worst in decision making pathologies by individuals and organizations.

For example, the months of July and August, 1914 present a rich tableaux of leaders who made mistaken assumptions about other states' intentions, capabilities, arts of war and politico–military staying power. Some heads of state and foreign ministers were unfamiliar with their own country's war plans and their implications for crisis management. In lieu of intelligence, stereotypical thinking about national character and military dispositions was available to take up the slack ("the Frenchman cannot be a very effective fighter, his voice is too high"). Added to this was the uncertainty about alliance cohesion on the part of the Triple Alliance and the Triple Entente: each state or empire had its own priorities, in policy and in strategy, and these priorities could not be synchronized under the time pressure between Sarajevo and the guns of August.

In a crisis involving two nuclear armed states with the capability for second strike retaliation, time pressure becomes nerve shattering. The evidence from studies of the Cuban missile crisis of 1962 shows that American and Soviet leaders operated under high personal stress and strained group decision making throughout the thirteen days that were required for the crisis to run its course. U.S. officials at one point wondered whether Soviet Premier Nikita Khrushchev had actually been the victim of a coup and replaced by a hard line Politburo coalition more determined for war. And the "known unknowns" as Donald Rumsfeld might have said are, in retrospect, equally discouraging for optimists about nuclear crisis management.

One of these "known unknowns" was whether the Soviets had deployed any nuclear capable delivery systems in Cuba in addition to the MRBM and IRBM launchers that provoked the crisis. U.S. officials at the time assumed not, but later historians determined otherwise. Nuclear capable surface-to-surface short range missiles were deployed with Soviet ground forces in Cuba, unknown to U.S. intelligence at the time. And Soviet ground force commanders, in the event of a U.S. military invasion, were presumably authorized to use nuclear capable missiles in self defense. The result of this "known unknown" could have been World War III, as a U.S. nuclear retaliation against Soviet nuclear first use in, or near, Cuba led to further escalation.

Preventive war or attack differs from preemption, nuclear or otherwise. Preventive war is anticipatory of a possible future attack, but not an inevitable one. Israel's attack on Iraq's nuclear reactor at Osirak in 1981 was motivated by Tel Aviv's concerns about what Saddam Hussein might do, should he acquire nuclear weapons at a future time. On the other hand, George W. Bush's attack on Iraq in 2003 was, if we take the President at his word, preemptive. Iraq was thought to have chemical and biological weapons in its possession by United States and other intelligence services, and its continuing interest in developing nuclear weapons was assumed on the basis of Saddam Hussein's prior stonewalling of UN international inspectors.

Case studies of military decision making lend themselves to conflicting interpretations. Two kinds of interpretations overlap: those of the policy makers and advisors who participated in the decision; and, second, those of academic or other observers of those decisions. Observers have the advantage of hindsight and distance from the actual events; insiders, the feel for the pressures experienced by those who had to act with incomplete information. For example, in the case of the Bush administration invasion of Iraq in 2003, the decision appears unwise in retrospect on account of the failure to find any weapons of mass destruction in Iraq. In addition, the botched occupation following the end of the active combat phase on May 1,

2003 casts additional retrospective doubt on the validity of the entire U.S. strategy and policy.

On the other hand, Bush policy makers were leaning forward into the decision, not backward against the harsh verdict of history. They did interpret some intelligence with a preconceived bias, for which they paid a significant cost in public credibility. However, all administrations do that: separating the "facts" of intelligence collection and analysis from the "interpretations" placed upon it by policy makers and military advisors is virtually impossible. An interesting aspect of the Bush II view of Iraq was that it was conditioned by the retrospective appraisal of the events of 9/11. Iraq was one front on the "war on terror" and Saddam Hussein might slip chemical or biological weapons to terrorists (or nuclear weapons, once he had them). Thus, by wrapping Iraq around the "war on terror" like a double helix, Bush, Cheney and their advisors perceived a pattern of strategic cooperation between Iraq and al-Qaeda where none really existed.

In reaction to the preceding critique, the George W. Bush administration might have responded that its war against Iraq was not preemptive, but rather preventive. It was to prevent Saddam Hussein from acquiring nuclear weapons in the future that he might use against Israel or give to terrorists. This justification might have merit if the Bush administration had not insisted that the danger posed by Iraq's WMD was *imminent*: that justification requires a case for preemptive, not preventive, war. The same problem applies to the Bush national security strategy that defends preemption as a necessary tool for policy makers and commanders under some circumstances. Few experienced policy planners or military analysts would argue the point, but Bush usage of "preemption" often elided into "preventive" war and vice versa.[30]

First use and first strike

The Cuban missile crisis provides an interesting overture for the second part of the problem of terminology related to nuclear first use: the distinction between nuclear "first use" and "first strike." Canonical Cold War usage referred to a nuclear first strike as an attack involving missiles or bombers of intercontinental range. Theater or shorter range attacks were usually described as first use.

---

[30] In this discussion, we are not entering into the debate whether the Bush administration purposely misrepresented the intelligence pertinent to the invasion of Iraq in 2003. That's a different topic and there is already a large literature about that subject. Sensible commentary on this point appears in Thomas E. Ricks, *Fiasco: The American Military Adventure in Iraq* (New York: The Penguin Press, 2006), esp. pp. 52–57.

However, this distinction was somewhat muddied by the overlap between geography, alliance membership and technology. An example is provided by the Soviet and then NATO deployment of Intermediate Nuclear Forces (INF) during the 1970s and 1980s before they were disarmed by treaty in 1987.

NATO ground launched ballistic missiles and ground and sea launched cruise missiles deployed in Europe were capable not only of striking targets within Eastern Europe, but also within Russia itself. Therefore, whereas NATO viewed its "572" deployments as offsetting capabilities in response to the Soviets' SS-20 ground launched ballistic missiles, Soviet military planners saw the NATO deployments as an escalation going beyond a symmetrical response to the Soviet initiative. One reason for this Soviet perception of NATO's intentions was the capability of U.S. Pershing II ballistic missiles to reach sensitive military and command targets in the western Soviet Union within minutes. Pessimistic Soviet military analysts might have interpreted the P-2s as a first strike weapon, intended to neutralize or obviate a Soviet retaliation following a NATO nuclear first use.

Further complicating the situation with respect to INF deployments was the two-way connection between INF and the ladder of escalation. Looking downward, INF were connected to the conventional forces deployed in Europe by both the NATO and Warsaw Treaty organization. Looking upward, INF were connected to the strategic nuclear deterrents of both the Americans and Soviets (and, with more uncertainty, to the British and French national nuclear forces, the latter conditionally available to NATO but solely under French determination). Thus the "intermediate" character of INF rested only on the technical dimensions of their range and probable destructive power. But the political "range" of INF capabilities were more problematical.

INF for the Soviets threatened to create a seamless preemptive theater war fighting capability in Europe that would, if put into effect, impose a military defeat or stalemate on NATO while at the same deterring American from escalating the conflict into a global nuclear war. INF for the Americans, from the Soviet perspective, threatened to undo this Soviet plan for "decoupling" NATO theater from American "strategic" nuclear forces by raising the stakes and risks of any "theater" nuclear first use. However, the U.S. and NATO "572" deployments could also raise the risks for NATO. Soviet war planners might decide that they had to attack the NATO INF immediately upon the outbreak of any large scale war, conventional or nuclear. Thus the INF could serve not as a firebreak between theater and global nuclear war, but as a detonator of the entire American and Soviet nuclear arsenals by inadvertence, or by design. Instead of contributing to a separation of conventional from nuclear war in Europe, or creating a firebreak between theater and strategic nuclear war, INF could expedite the leap from nuclear first use into total war.

In short, INF deployments were soon realized by both the Soviets and NATO to have created a zone of uncertainty with respect to deterrence and the control of escalation that was unacceptable. The walk from "first use" to "first strike" was too quick and too ambiguous for diplomats and war planners to sort out in the exigent circumstances of the "fog of war." It was problematical enough to maintain any clear firebreak between tactical and strategic weapons once the nuclear threshold had been crossed – a distinction that the Soviets, as a matter of practice, disavowed, although they were well prepared for tactical nuclear first use apart from ordering a nuclear first strike by their long range forces.

The case of INF in Europe shows how the line between first strike and first use is as much a matter of arbitrary definition as it is a reliable guide to military effectiveness or deterrence credibility. If nuclear weapons of shorter range and lesser yields were capable of being used with the surgical precision of conventional weapons, then shorter range and lower yield nuclear weapons would be stronger candidates for preemption and first use or first strike missions. However, the advent of sanitized nuclear weapons, comparable in their collateral damage to conventional means, is not imminent, and ironically, not judged to be desirable by politicians or military planners. Nuclear weapons derive their deterrent effects from their "awfulness:" their capability to destroy not just military targets, but societies and economies on a large scale, in a historically unprecedented short period of time. Even the most obtuse politician is thus pushed backward from candidate scenarios of "victory" on offer from briefers on first use or first strike.

A second implication of the results in the preceding analysis has to do with the issue of "no first use" as a declaratory or operational policy for American or other nuclear forces. No first use of nuclear weapons is an ethically admirable, and politically desirable, declaratory policy for states to have. However, it is highly conditional on circumstances and scenario dependent as to its effectiveness. NATO found a no first use policy inexpedient during the Cold War, on account of the presumed inferiority of NATO's conventional forces compared to those of the Soviet Union and Warsaw Pact deployed in Europe. Russia now finds a no first use declaratory policy unpropitious for the same reason: the decrepit character of its conventional forces, compared to those of the United States and NATO or to Soviet forces of the late Cold War.[31] NATO considers that its relationship with Russia is officially one of openness to security cooperation, including the possibility of post-New

---

[31] Dale R. Herspring, "Putin and Military Reform," Ch. 8 in Herspring, ed., *Putin's Russia: Past Imperfect, Future Uncertain* (Lanham, Md.: Rowman and Littlefield, 2007), Third Edition, pp. 173–194.

START reductions in Russian and American tactical nuclear weapons deployed in Europe. Nevertheless, an expert panel working up a new strategic concept for NATO reported in May, 2010 that:

> As long as nuclear weapons exist, NATO should continue to maintain secure and reliable nuclear forces, with widely shared responsibility for deployment and operational support, at the minimum level required by the prevailing security environment. Any change in this policy, including in the geographic distribution of NATO nuclear deployments in Europe, should be made, as with other major decisions, by the Alliance as a whole.[32]

It is argued that no first use doctrines are sometimes dysfunctional for deterrence, especially for the deterrent umbrella that the United States might want to extend to allies. As a case in point, the United States might want some states in the Middle East or Asia to be deterred from attacking regional American allies (Taiwan, Japan, Israel, Iraq) with conventional forces or with weapons of mass destruction other than nuclear. The credible threat of nuclear first use against such adventurism might give pause to aggressors who would otherwise be willing to gamble on U.S. restraint. For example, U.S. negotiators apparently informed Iraqi leader Saddam Hussein in 1991, prior to the outbreak of Operation Desert Storm, that any Iraqi use of chemical or biological weapons would put all American options on the table, including the possible first use of nuclear weapons. On the other hand, this case might be interpreted not as a case of deterrence, but as an instance of escalation control for the management of a conflict that U.S. officials and Iraqis knew to be inevitable.

Extended deterrence does have the value of providing a U.S. nuclear umbrella over states in Europe or Asia that might have deployed their own nuclear weapons in lieu of American protection. On the other hand, demonstrating that extended deterrence has "worked" because of American nuclear weapons, as opposed to other assets, is a more difficult brief now than it would have been during the Cold War. In conventional warfare, the United States, in the first decade of the twenty-first century, was unarguably superior to any other state as a military power with global reach.[33] The case that nuclear umbrellas, as opposed to conventional raincoats, are necessary

---

[32] *NATO 2020: Assured Security; Dynamic Engagement, Analysis and Recommendations of the Group of Experts on a New Strategic Concept for NATO* (Brussels: North Atlantic Treaty Organization, May 17, 2010), p. 44. See also: "NATO Mission Statement Supports Retaining Tactical Nukes," *Global Security Newswire*, May 17, 2010, http://gsn.nti.org/gsn/nw_20100517_8029.php.

[33] Stephen M. Walt, *Taming American Power: The Global Response to U.S. Primacy* (New York: W.W. Norton, 2005), pp. 33–36.

for the protection of allies against threats *other than nuclear coercion or attack* is weaker now than hitherto. As nuclear nonproliferation expert George Perkovich has noted:

> Postulating the utility of nuclear weapons to deter anything less than threats to national survival may induce states to count on these weapons to prevent or halt crises in which nuclear weapons should or would not be used. It could even embolden allies to take actions that could start or intensify crises, counting on U.S. nuclear weapons to back them up.[34]

A third aspect of this discussion is what to do with some 200 U.S. sub-strategic nuclear weapons deployed in four European countries and in Turkey. One school of thought regards these weapons as vital reminders of NATO's commitment to the defense of its expanded post-Cold War membership. From this perspective, U.S. European-based nuclear weapons reinforce the "coupling" of United States and allied European political and military capabilities by the creation of shared risk, including the risk of nuclear escalation in, and beyond, Europe. Responding to proposals for the removal of U.S. nuclear weapons from Germany, and perhaps from Europe altogether, some expert commentators warned against possible repercussions for the credibility of NATO strategy and policy:

> Poland and the Baltic states in particular are likely to argue with merit that a withdrawal of nuclear weapons from Europe would constitute a material change to those commitments [to their defense], and to NATO's mutual defense guarantee, "Article V", as they understood it. They will be particularly worried that the security of the United States is being decoupled from the security of Europe – the new NATO countries still trust the United States more than their west European counterparts.[35]

On the other hand, the European political alignments of the twenty-first century are not those of the Cold War. The preparedness for an entire spectrum of nuclear wars, from local and regional to global conflicts, is neither necessary for credible deterrence nor responsive to the realistic threats facing modern Europe. Instead of former Soviet operational maneuver groups pushing

---

[34] George Perkovich, *Nuclear Weapons in Germany: Broaden and Deepen the Debate* (Washington, D.C.: Carnegie Endowment for International Peace, February 2010), p. 6.

[35] Franklin Miller, George Robertson and Kori Schake, *Germany Opens Pandora's Box* (London: Centre for European Reform), February, 2010, www.cer.org.uk. See also: David J. Kramer, "U.S. abandoning Russia's neighbors," *Washington Post*, May 15, 2010, in *Johnson's Russia List* 2010 – #96, May 17, 2010, davidjohnson@starpower.net.

westward with massive infantry and armored thrusts, NATO members now face the threats posted by possible Russian (or other) coercive diplomacy, including cyber-bullying, resource strangulation, brooding military maneuvers, and instigating diplomatic demarches intended to throw NATO cohesion off balance.[36] If, against political odds, war does happen, U.S. nuclear weapons deployed on European soil may invite preemptive attacks on themselves instead of deterring escalation to nuclear first use.[37]

As an alternative to a declaratory policy of nuclear first use, the nuclear powers might consider the doctrine of "defensive last resort." Defensive last resort is one step less rigid than nuclear first use. A doctrine of last resort (presumably defensive in intent) was adopted by NATO in 1991 and, as a declaratory policy, it is more suited to the realities of operational policy and military practice. Under a doctrine of defensive last resort, the first use of nuclear weapons is not precluded, but it is also not encouraged as an early step on the ladder of escalation. As explained by the authors of an important study on nuclear arms control:

> To recognize the possibility that in some future defense against aggression the use of the nuclear weapon could unexpectedly become the only alternative to an even worse disaster is not to encourage reliance by planners on any such action, nor does it support any doctrine of early use. A doctrine of defensive last resort is fully consistent with a continuing American effort to sustain the worldwide tradition of nonuse.[38]

The preceding point is reinforced by the blurred line between nuclear first use and first strike already noted in this discussion, and by the unhealthy dependency of current and possible future nuclear states on prompt launch and high alert (i.e., hair triggers) in order to guarantee the survivability and retaliatory credibility of their nuclear forces.[39]

---

[36] Pertinent examples appear in Perkovich, *Nuclear Weapons in Germany*, passim.

[37] For a superior treatment of this subject, especially with regard to the trade-offs between preserving the credibility of U.S. extended deterrence for Europe, on one hand, and achieving U.S. and NATO nonproliferation and arms reduction goals, on the other, see Malcolm Chalmers and Simon Lunn, *NATO's Tactical Nuclear Dilemma* (London: Royal United Services Institute, Occasional Paper, March 2010), www.rusi.org.

[38] McGeorge Bundy, William J. Crowe, Jr. and Sidney D. Drell, *Reducing Nuclear Danger: The Road Away from the Brink* (New York: Council on Foreign Relations, 1993), p. 85.

[39] George H. Quester makes the important point that the first use of nuclear weapons since Nagasaki might be shrouded in ambiguity as to whether a state or other actor was responsible, whether the attack was accidental or deliberate, or other factors. See Quester, *Nuclear First Strike*, pp. 24–43 and passim.

Conclusion

Notwithstanding the bonhomie of the Lisbon summit of 2010 and the officially declared nonthreatening character of NATO and Russian winds blowing eastward and westerly, important military issues in Europe related to nuclear weapons remain to be resolved. Most important, United States, NATO or even Russian declaratory policies, let alone extensive debates, about nuclear first use or first strike are unhelpful as matters of public diplomacy. As matters of military credibility or deterrence stability, they are even worse. There is little to be gained, and much potentially to be lost, by front-ending nuclear weapons onto undisciplined "what if" policy discussions. In an exceptional case that requires serious consideration of nuclear first use, or the threat of same, leaders can rise to the occasion without having already mortgaged their reputation for seriousness and sanity.

The threat of nuclear first use against terrorists with WMD or states that harbor them is hardly likely to dissuade the terrorists, although it may inhibit other states from providing comparable support to dangerous malcontents. However, apocalyptic terrorists might actually welcome a preventive nuclear attack on their headquarters and storage sites, providing them with martyrdom, and inflaming much of the rest of the world against American ideals and policies. Nuclear weapons are neither the obvious first choice for suppression of nonstate actors by preemptive military attacks, nor the expedient solution to a problem that is best resolved by improved intelligence, better international cooperation in counterterror operations and lethal non-nuclear munitions. In addition, as former U.S. Senator and nuclear policy expert Sam Nunn has noted, Europe's dated Cold War nuclear infrastructure actually invites security lapses, appealing to lustful terrorists seeking bombs:

> If we don't address this issue with urgency, we may wake up one day to a 1972 Munich-Olympics scenario, with a masked terrorist waving a gun outside of a nuclear warhead bunker somewhere in Europe. This time the hostages could be millions of people living close by.[40]

---

[40] Sam Nunn, "NATO, Nuclear Security and the Terrorist Threat," *New York Times*, November 16, 2010, http://www.nytimes.com/2010/11/17/opinion/17/iht-ednunn.html.

8

Minimum deterrence and missile defenses: Congruent paths or competitive designs?

Introduction

The endgame debates over New START ratification within U.S. and Russian policy making circles suggested that a pause might be advisable before proceeding to more ambitious reductions in offensive nuclear weapons. Regardless of this, the Obama administration has directed the Department of Defense to study the possibility of further reductions in U.S. and Russian strategic nuclear weapons, possibly well below already agreed New START levels.[1] Russia's incentive structure for post-New START denuking is a complicated mixture of its international threat perceptions, domestic policy needs, and expectations for the reach of the "reset" in Russian–American relations.

First, a perception of continued, post-New START strategic nuclear equivalence between the United States and Russia provides a security glacis, behind which Russia can politically afford to engage in military reform of its

---

[1] See "Pentagon Studying Additional Nuke Reductions," *Global Security Newswire,* March 23, 2011, *http://gsn.nti.org/siteservices/print_friendly.php*, and Desmond Butler, "US reviewing nuclear arsenal with eye to new cuts," Associated Press, March 23, 2011, in *Johnson's Russia List 2011* – #53, March 23, 2011, davidjohnson@starpower.net.

conventional forces. Second, stasis at New START levels without further bilateral reductions will vitiate American and Russian claims to leadership in nonproliferation and nuclear arms control worldwide. Third, the issue of offensive nuclear arms reductions as between the United States and Russia has a flip side – emerging technologies for missile defense. The Obama administration plan for future ballistic missile defense (BMD) deployments in Europe, although less unacceptable to the Kremlin than the earlier George W. Bush proposal, caused Russian angst during the debate over New START and creates additional uncertainty about Russia's future compliance with that agreement.[2] On the other hand, NATO and Russia in March, 2011 began high level talks on possible cooperation in developing and operating a European regional missile defense system.[3] Can a march downward to minimum deterrence, based on post-New START reductions in offensive long range nuclear weapons, coexist peacefully with joint or singular missile defense deployments in Europe by NATO and Russia?

Perspective on these topics is provided in the following discussion, as below. First, some of the broader politico–military and arms control contexts for the viability of any minimum nuclear deterrence regime are specified. Second, minimum deterrence is located within a spectrum of broader deterrence/arms control options. Third, the analysis considers whether missile defenses if successfully deployed would threaten the viability of a minimum deterrence regime.

Why minimum deterrence?

The idea of minimum deterrence has caught fire among civilian and military policy analysts and other close students of nuclear arms control.[4] Minimum

---

[2] Linking agreements on U.S.–Russian offensive nuclear force reductions to prospective deployments of European missile defenses is mistaken and "fundamentally wrong" according to prominent Russian missile designer Yuri Solomonov. See "Russian missile designer criticizes New START dependence on missile defense," *Interfax*, March 17, 2011, in *Johnson's Russia List* 2011 – #50, March 18, 2011, davidjohnson@starpower.net.

[3] See: "Missile defense is a 'sincerity test' for NATO – Lavrov," www.russiatoday.com, March 23, 2011, in *Johnson's Russia List* 2011 – #53, March 23, 2011, davidjohnson@starpower.net; and "Russian deputy foreign minister seeks specific missile defence deal with the US," *BBC Monitoring*, Ekho Moskvy Radio, March 17, 2011, in *Johnson's Russia List* 2011 – #50, March 18, 2011, davidjohnson@starpower.net.

[4] For important arguments and pertinent citations in recent studies, see: James Wood Forsyth Jr., B. Chance Saltzman and Gary Schaub Jr., "Minimum Deterrence and Its Critics," *Strategic Studies Quarterly*, No. 4 (Winter, 2010), pp. 3–12; Forsyth, Saltzman and Schaub, "Remembrance of Things Past: The Enduring Value of Nuclear Weapons," *Strategic Studies Quarterly*, No. 1 (Spring, 2010), pp. 74–89; and Stephen M. Walt, "All the nukes you can use," foreignpolicy.com, May 24, 2010, http://walt.foreignpolicy.com/posts/2010/05/24/all_the_nukes_that_you_can_use.

deterrence might appeal as an acceptable alternative to the more utopian construct of nuclear abolition, endorsed in principle by U.S. President Barack Obama and a number of leading former policy makers and military commanders.[5] Minimum deterrence might also be acceptable to military planners who want to maintain a viable U.S. nuclear deterrent at an acceptable cost.[6] In addition, experts on nuclear nonproliferation might favor minimum deterrence as a way station toward multilateral nuclear arms reductions and further measures of cooperative threat reduction among nuclear weapons states as well as nuclear-threshold or nuclear aspiring powers.[7]

However, discussion of minimum deterrence can bring participants into the land of mystery and confusion, unless the discussion is disciplined by political and military-strategic clarity. A nuclear deterrent force can be described as "minimum" or "maximum" depending upon the security dilemmas facing various states, including their expectations about probable opponents' security objectives, military capabilities and decision making styles. Pakistan, Britain and Israel all are regarded as nuclear weapons states, but their

---

[5] See, for example: George P. Shultz, William J. Perry, Henry A. Kissinger and Sam Nunn, "Toward a Nuclear-Free World," *Wall Street Journal*, January 15, 2008, p. A13. See also: Henry A. Kissinger, "Containing the fire of the gods," *International Herald Tribune*, February 6, 2009, http://www.iht.com/articles/2009/02/06/opinion/edkissinger.php; Press Association, "David Miliband sets out six-point plan to rid world of nuclear weapons," guardian.co.uk, February 4, 2009, http://www.guardian.co.uk/politics/2009/feb/04/miliband-nuclear-weapons; and Jennifer Loven, "Obama outlines sweeping goal of nuclear-free world," *Associated Press*, April 5, 2009, in *Johnson's Russia List* 2009 – #66, April 5–6 2009, davidjohnson@starpower.net. For insightful commentary on this topic, see Thomas C. Schelling, "A world without nuclear weapons?" Daedalus (Fall, 2009), No. 4, pp. 124–129; Jonathan Schell, *The Seventh Decade: The New Shape of Nuclear Danger* (New York: Henry Holt and Co., 2007), esp. pp. 201–223; Lawrence Freedman, "Eliminators, Marginalists, and the Politics of Disarmament," Ch. 4 in John Baylis and Robert O'Neill, eds., *Alternative Nuclear Futures: The Role of Nuclear Weapons in the Post-Cold War World* (Oxford: Oxford University Press, 2000), pp. 56–69; and Colin S. Gray, The Second Nuclear Age (Boulder, Colo.: Lynne Rienner, 1999), Ch. 4, esp. pp. 82–85.

[6] Minimum deterrence has, in fact, a considerable pedigree, dating back to some of the earliest U.S. debates on nuclear strategy and deterrence. "Minimum deterrent" strategies have variations and are sometimes referred to as "deterrence only" or "finite deterrence" strategies. See Herman Kahn, *On Thermonuclear War*, Second Edition (New York: The Free Press, 1969), pp. 7–13; and Kahn, *On Escalation: Metaphors and Scenarios* (New York: Frederick A. Praeger, 1965), pp. 281–284. See also: John Baylis, "Nuclear Weapons, Prudence, and Morality: The Search for a 'Third Way'," Ch. 5 in Baylis and O'Neill, eds., *Alternative Nuclear Futures*, pp. 70–86 inclusive, esp. pp. 78–81.

[7] An expert assessment in 1999 concluded that nuclear abolition was impractical of realization, leaving the practical question whether the United States could or should reduce its arsenal to hundreds of nuclear weapons any time in the next two or three decades. See: Center for Nonproliferation Research – National Defense University, and Center for Global Security Research, Lawrence Livermore National Laboratory, *U.S. Nuclear Policy in the 21st Century: A Fresh Look at National Strategy and Requirements* (Washington: U.S. Government Printing Office, 1998), esp. pp. 3.15–3.18.

perceived security dilemmas, expectations about deterrence requirements, and decision making patterns vary markedly. Minimum deterrence is not one remedy that fits all states, but a conceptual framework that could induce helpful expectations about deterrence stability and security cooperation, given favorable political winds. From the same perspective, the "adequacy" of a minimum or larger deterrent cannot be defined by numbers of weapons alone, but by the political and military-strategic context within which they might be used – for deterrence, or otherwise.

Assumptions can go wild as deuces in discussions of nuclear deterrence, since we thankfully have no historical examples of a two sided conflict with nuclear weapons. The following assumptions guide further discussion here.[8] First, the primary purpose of nuclear weapons is to deter nuclear attack or nuclear coercion. The avoidance of war by means of deterrence is the ultimate rationale for nuclear weapons. Second, deterrence is a subjective, not an objective, construct. Deterrence is in the eye of the state being deterred. The recipient of a deterrent threat, not the threatener, gets the final vote on whether deterrence has "worked."[9] Third, a failure of deterrence that results in the first nuclear attack of the twenty-first century will be an historical page turner and political "game changer" that reboots some prior assumptions about politics and military strategy. Reactions to the first nuclear weapon fired in anger since Nagasaki might be dominated by the feeling that the "nuclear taboo" had been irretrievably shattered. On the other hand, there is the opposite possibility that such an event might sober leaders and promote greater determination toward international arms control and denuclearization: in other words, leaders and publics might run away from religion or embrace it all the more fervently.

A fourth assumption is that there is not necessarily any single number of deployed weapons that can guarantee a state that its deterrent is proof against intimidation or first strike destruction. Although the numbers of weapons deployed might matter in a political confrontation between nuclear armed states, the numbers are meaningful only within a larger context of decisions about the deployment of nuclear capable launchers, the survivability

---

[8] For assessments of deterrence before and after the Cold War, see: Michael Krepon, *Better Safe than Sorry: The Ironies of Living with the Bomb* (Stanford, Calif.: Stanford University Press, 2009); Patrick M. Morgan, *Deterrence Now* (Cambridge: Cambridge University Press, 2003); Lawrence Freedman, *The Evolution of Nuclear Strategy* (New York: Palgrave Macmillan, Third Edition, 2003); Colin S. Gray, *The Second Nuclear Age* (Boulder, Colo.: Lynne Rienner, 1999); Gray, *Modern Strategy* (Oxford: Oxford University Press, 1999), Ch. 11–12; Keith B. Payne, *Deterrence in the Second Nuclear Age* (Lexington, Ky.: University Press of Kentucky, 1996); and Robert Jervis, *The Meaning of the Nuclear Revolution: Statecraft and the Prospect of Armageddon* (Ithaca, N.Y.: Cornell University Press, 1989).

[9] On this point, see Morgan, *Deterrence Now*, p. 164.

of launchers and command-control systems, and extant protocols for dealing with nuclear warning and response. States are not automatons, and military forces are living social organisms as well as collections of weapons and procedures. There is no guarantee that any number of nuclear weapons deployed in any configuration will be supported in extremis by rational or sensible decision making. Thus the study of civil–military relations in nuclear armed states becomes all the more important as more states seek or acquire nuclear forces, notwithstanding how undesirable the further spread of these weapons might be.

A fifth assumption is that nuclear weapons deployments and deterrence based on nuclear forces do not take place in an historical or military–technical vacuum. We are in the information age; nuclear weapons were products of the industrial age and its emphasis on mass destruction as the acme of strategy and military art. The information age privileges precision-targeted, long range and stealthy conventional weapons and C4ISR systems (command, control, communications, computers, intelligence, surveillance and reconnaissance) that enable network-centric warfare across a spectrum of non-nuclear conflicts. In this context, nuclear weapons appear as atavistic troglodytes that belong in Indiana Jones's museum: good for mass killing but not for effects based operations, as they say in the Pentagon. However, nuclear weapons, however retro from the standpoint of high end, information-based conventional forces, retain a stubborn appeal to states for reasons of deterrence and dissuasion, prestige, swaggering, and head butting, even as they raise the prospect of scoring an "own goal" in the process of actual use. As Lawrence Freedman has noted, it remains the case in the present century that, although there is no such thing as a "nuclear strategy" per se, military strategists must take nuclear weapons into account.[10]

Given these assumptions, we proceed to discuss and analyze the concept of minimum deterrence within the context of present and foreseeable national security and nuclear arms control issues. The elements of that context are: (1), a recently ratified New START agreement between the United States and Russia, scheduled to enter into force on February 5, 2011;[11] (2), the possibility

---

[10] Lawrence Freedman, *The Evolution of Nuclear Strategy*, Third Edition (New York: Palgrave Macmillan, 2003), p. 464.

[11] The text of the New START treaty appears in *Treaty between the United States of America and the Russian Federation on Measures for the Further Reduction and Limitation of Strategic Offensive Arms* (Washington, D.C.: U.S. Department of State, April 8, 2010), http://www.state.gov/documents/organization/140035.pdf. See also: "Lavrov, Clinton to bring New START into force on February 5," *RIA Novosti*, February 1, 2011, http://en.rian.ru/military_news/20110201/162410687.html. Indispensable as a resource is Pavel Podvig, "New START Treaty in numbers," from his blog, *Russian strategic nuclear forces*, April 9, 2010, http://russianforces.org/blog/2010/03/new_start_treaty_in_numbers.shtml. Contrasting appraisals of New START appear in: Steven Pifer, "New

of further, post-New START reductions in U.S. and Russian strategic nuclear weapons, perhaps leading to multilateral force reductions among the P-5 and other existing nuclear weapons states; and, (3), President Obama's agenda for further denuclearization and arms limitation. The last point involves American and other outliers' willingness to ratify the Comprehensive Test Ban Treaty (CTBT); passage of an international Fissile Materials Cutoff Treaty (FMCT); and, third, NATO–Russian security cooperation with respect to the limitation of nonstrategic nuclear weapons deployments in Europe, the shared development and-or deployment of theater missile defenses in Europe, and the transformation of political expectations in Europe to include an Atlantic-to-Urals regime of assumed security community (defined as the absence of any expectation of intramural warfare among its members, as now exists among member states of NATO).

On the possibility of redefining security space in Europe including western Russia, the conclusion of New START does not necessarily provide an assured glide path toward a preferred destination, either for NATO or for Russia. Obstacles standing in the way of further progress toward denuclearization and demilitarization in Europe include: (1), Russian doubts about U.S. intentions with respect to European and global missile defenses; (2), disagreements among NATO members about the trustworthiness of Russia as a security partner; and, (3), the asymmetrical military positions of NATO and Russia, with respect to their relative dependencies on conventional or nuclear weapons. Doubtless, domestic politics in the member states of NATO and within Russia also play into disagreements about each of these three issue areas.

Regarding the first issue, missile defenses, the Obama administration revised the George W. Bush plan for missile defenses in Europe for both political and military-technical reasons. The revised plan, according to Defense Secretary Robert Gates and experts from the Pentagon's missile defense agency, would permit quicker deployment of more reliable antimissile technologies, on sea based platforms as well as land sites, than called for in the original Bush program.[12] In political terms, the revised Obama ballistic

---

START: Good News for U.S. Security," *Arms Control Today*, May 2010, http://www.armscontrol. org/print/4209; Keith B. Payne, "Evaluating the U.S.–Russian Nuclear Deal," *Wall Street Journal*, April 8, 2010, in *Johnson's Russia List* 2010 – #69, April 8, 2010, davidjohnson@starpower.net; Jonathan Schell, "Nuclear balance of terror must end," CNN, April 8, 2010, in *Johnson's Russia List* 2010 – #69, April 8, 2010, davidjohnson@starpower.net; and Alexander Golts, "An Illusory New START," *Moscow Times*, March 30, 2010, in *Johnson's Russia List* 2010 – #62, March 30, 2010, davidjohnson@starpower.net.

[12] Robert M. Gates, "A Better Missile Defense for a Safer Europe," *New York Times*, September 19, 2009, http://www.nytimes.com/2009/09/20/opinion/20gates.html. The Obama Phased Adaptive Approach to missile defense will retain and improve some technologies deployed by the George W. Bush administration, but shift emphasis to other interceptors, supported by improved battle

missile defense (BMD) plan was intended to mollify Russian concerns about the possibility that American missile defenses deployed in Europe would nullify Russia's assured nuclear second strike capability.[13] Russian fears on this point can be traced all the way back to the Cold War and U.S. President Ronald Reagan's proposal for a nationwide Strategic Defense Initiative to render ballistic missiles obsolete. The technical immaturity of missile defenses relative to offenses is less impressive for Russian pessimists than are their doubts about American and NATO intentions.[14] Nevertheless, in the aftermath of the New START agreement, NATO and Russia have agreed to

---

management – command-control-communications (BMC3) systems and launch detection and tracking. See: Tom Z. Collina, "Missile Defense Test a 'Success': Pentagon," *Arms Control Today*, May 2011, Arms Control Association, http://www.armscontrol.org/print/4860, downloaded May 6, 2011; *Unclassified Statement* of Lieutenant General Patrick J. O'Reilly, USA, Director, Missile Defense Agency, Before the House Armed Services Committee (Washington, D.C.: U.S. House of Representatives, House Armed Services Committee, October 1, 2009). Early assessments of the revised Obama missile defense plan include: George Friedman, "The BMD Decision and the Global System," Stratfor.com, September 21, 2009, in *Johnson's Russia List* 2009 – #175, September 22, 2009, davidjohnson@starpower.net; Alexander Golts, "Calling Moscow's Bluff on Missile Defense," *Moscow Times*, September 22, 2009, in *Johnson's Russia List* 2009 – #175, September 22, 2009, davidjohnson@starpower.net; Alexander L. Pikayev, "For the Benefit of All," *Moscow Times*, September 21, 2009, in *Johnson's Russia List* 2009 – #174, September 21, 2009, davidjohnson@starpower.net; and Strobe Talbott, "A better base for cutting nuclear weapons," *Financial Times*, September 21, 2009, in *Johnson's Russia List* 2009 – #174, September 21, 2009, davidjohnson@starpower.net.

[13] See "U.S. launches new missile defense program for Europe," *Associated Press*, March 1, 2011, in *Johnson's Russia List* 2011 – #38, March 2, 2011, davidjohnson@starpower.net, and Alexander Gabuyev, "Development of the European ballistic missile defense system is under way," *Kommersant*, March 3, 2011, in *Johnson's Russia List* 2011 – #39, March 3, 2011, davidjohnson@starpower.net. A critical expert appraisal of the Obama missile defense plan appears in George N. Lewis and Theodore A. Postol, "A Flawed and Dangerous U.S. Missile Defense Plan," *Arms Control Today*, May 2010, http://www.armscontrol.org/print/4244. See also: William J. Broad and David E. Sanger, "Review Cites Flaws in U.S. Antimissile Program," *New York Times*, May 17, 2010, http://www.nytimes.com/2010/05/18/world/18missile.html. A more favorable expert assessment appears in Hans Binnendijk, "A Sensible Decision: A Wider Protective Umbrella," *Washington Times*, September 30, 2009, in *Johnson's Russia List* 2009 – #181, September 30, 2009, davidjohnson@starpower.net. Continuing Russian doubts are noted in "Russia Still Suspicious of U.S. Missile Defense Plans," *Reuters*, September 29, 2009, in *Johnson's Russia List* 2009 – #181, September 30, 2009, davidjohnson@starpower.net. For additional background, see Arms Control Association, *U.S. Missile Defense Programs at a Glance*, undated, http://www.armscontrol.org/print/4061, downloaded May 19, 2010.

[14] Thus, for example, Russia's permanent representative to NATO Dmitry Rogozin insisted that Russia must play an equal role with NATO in setting up any European missile defense system because its paramount interests "lie in the protection of our sovereignty and independence that are guaranteed by our strategic nuclear potential." Rogozin, cited in "Russia Wants To Be Nato's Equal Partner in Euro ABM – Envoy," *Interfax*, March 3, 2011, in *Johnson's Russia List* 2011 – #40, March 4, 2011, davidjohnson@starpower.net.

reboot discussions on cooperative missile defenses with respect to threats from Europe posed by outside nuclear forces, for example, Iran.[15]

Second, doubts held by some Russians about NATO's trustworthiness as a security partner are mirrored by some governments in NATO, especially those states that are party to future coercion by a Russia with a modernized and rebuilt military. These doubts about Russia within the ranks of NATO impact upon expectations for conventional or nuclear arms control in Europe. Doubts in both directions, East–West and West–East, also influence debates over further expansion in the membership of NATO. Russia's leadership has made it clear that the accession of Georgia or Ukraine to full membership in NATO would be troublesome to trans-Atlantic relations, and in the case of Ukraine, probably destructive of Obama's reset, and more. NATO's perceived need for Russian support for the alliance's efforts in Afghanistan is also hostage to Russia's opposition to further NATO expansion.

A third pertinent issue is the post-Cold War asymmetry between NATO and Russia with respect to their reliance on nuclear as compared to conventional military forces as the makeweights of their national security policies. Russia's weaker conventional forces relative to NATO's, place more burden on its strategic and nonstrategic nuclear weapons to cover a wider range of deterrent situations and conflicts. Conversely, the United States and NATO have advantages in information based conventional warfare, including bellwether technologies for precision deep strike, C4ISR and stealth, among others. Russia's greater reliance on nuclear weapons includes both its strategic and nonstrategic nuclear forces. This poses the risk for NATO that even an outbreak of conventional warfare across the fault lines of former Soviet security space could rapidly trip into tactical or theater nuclear war, supported by the brooding omnipresence of Russian and American strategic nuclear weapons. Granted, in the present political climate, such an event is not only improbable, but unimaginable. But Russian and other military planners get paid to develop contingency plans for worst cases as well as better ones.

Underlying all of the preceding discussion is the matter of choosing among paradigms for the interpretation of past and current security related activity and, as well, for anticipation of probable futures, however hazardous the enterprise.

---

[15] See, on this topic: Daniel Wagner and Diana Stellman, "The Prospects for Missile Defense Cooperation Between NATO and Russia," *Foreign Policy Journal*, February 10, 2011, www.foreignpolicyjournal.com, in *Johnson's Russia List* 2011 – #24, February 10, 2011, davidjohnson@starpower.net; Dr. Donald N. Jensen, "Disagreements over Missile Defense Threaten to Undermine new START Treaty," *Voice of America Russian Service*, February 9, 2011, www.voanews.com/Russian/news in *Johnson's Russia List* 2011 – #25, February 11, 2011,davidjohnson@starpower.net; and Vladimir Kuzmin, "From START to Euro-ABM," *Rossiyskaya Gazeta*, January 31, 2011, in Johnson's Russia List 2011 – #18, January 31–February 2, 2011, davidjohnson@starpower.net.

One simply cannot get away from the problem of paradigms and futures because, for example, a U.S. or other policy of minimum nuclear deterrence makes more sense in some kinds of worlds than in others. Alternative nuclear worlds can be imagined that include: (1), freezing the number of nuclear weapons states at the present (early 2011) number, and possibly reversing North Korean nuclear weapons status, with unsuccessful terrorist efforts to acquire or to use nuclear weapons; (2), continuing "slow rolling" nuclear proliferation among state actors, with unsuccessful terrorist efforts to acquire or to use nuclear weapons; (3), slowing nuclear weapons spread among state actors, with successful acquisition of nuclear materials by at least one terrorist group, although no actual use of nuclear weapons; (4), slowing the spread of nuclear weapons among state actors, with successful acquisition of nuclear materials by at least one terrorist group which uses them for nuclear blackmail without actual detonations; and, (5), rapid spreading of nuclear weapons among state actors, with successful acquisition of nuclear materials by at least one terrorist group which detonates an actual nuclear bomb in one city. Doubtless other possibilities can be imagined, but the point is clear.

In nuclear worlds 1 and 2, as above, one can still hypothesize that minimum deterrence could be a soft sell among some, although not all, of the existing nuclear weapons states. Even in world 3, where nuclear weapons spread among states remains slow and terrorists have acquired but not used nuclear weapons for blackmail or attacks, a briefing for minimum deterrence might hold the attention of an expert audience. But in worlds 4 or 5, slow or fast proliferation among states and actual terrorist uses of nuclear materials and-or nuclear weapons for blackmail or murder, minimum deterrence will be a hard sell for active duty policy makers and military planners, whatever its appeal to theorists.

Since political and military "futures" are indeterminate, policy makers have the opportunity to shape their environments toward constrained nuclear proliferation.[16] Factors favoring constrained nuclear weapons spread might include: (1), the ethical and moral inhibitions on the part of many governments and military professionals against the possession or use of nuclear weapons for deterrence or for warfare; (2), the possibility that sensible or rational decision makers will find other and less destructive military means to accomplish their political objectives, including advanced conventional weapons, alliances for

---

[16] See Stephen J. Blank, *Russia and Arms Control: Are There Opportunities for the Obama Administration?* (Carlisle, Pa.: Strategic Studies Institute, U.S. Army War College, March 2009), for an assessment of possible areas of U.S.–Russian cooperation and pertinent obstacles. Important trends in Russian security and defense policy are traced in Olga Oliker, Keith Crane, Lowell H. Schwartz, and Catherine Yusupov, *Russian Foreign Policy: Sources and Implications* (Santa Monica, Calif.: RAND Corporation, 2009), Ch. 5, esp. pp. 162–174.

extended deterrence protection by existing nuclear powers, or modification of their political objectives for the purpose of war avoidance; and, (3), the inertial effect of a presumed "nuclear taboo" in existence since Nagasaki, and the related uncertainty of a new world following the first nuclear attack in the twenty-first century.[17] There are, in symbolism as well as in substance, no such things as "small" nuclear wars or nuclear attacks.[18]

Of course, the contrasting forces that might favor additional nuclear weapons spread among state actors, with possible spillover into the hands of terrorists, are recognized by many experts. States see nuclear weapons as deterrents against nuclear coercion or attack from other states. Nuclear weapons are thought by some states to confer prestige and to provide a cost-effective entry into the ranks of major military powers. Other states might see nuclear weapons as a "last ditch upper of the ante" against a catastrophic defeat in a conventional war.[19] Nuclear weapons are also used for diplomatic swaggering and posturing in order to buff the image of states whose military credibility might otherwise be doubted by onlookers and possible adversaries. Finally, nuclear weapons can serve a variety of domestic policy needs for states and regimes, including the appeasement of powerful military or nuclear industry groups and hawkish parliamentary factions.

One critical indicator of which "world" we're headed for is the fate of the existing nuclear nonproliferation regime. In this writer's judgment, that regime of international institutions, procedures and consultations has performed admirably in the past and with more success than pessimists, including former U.S. presidents and prominent nuclear weapons scientists, had anticipated. In some ways it performed too well, leading to complacency in some quarters that nuclear weapons in the twenty-first century will continue to spread slowly, if at all. We agree with Kenneth Waltz's argument (to a point) that the existence of survivable nuclear forces can induce caution on the part of otherwise attack prone or brinkmanship oriented political leaders and their military advisors.[20] Cold War experience supports this argument. On the other hand, Scott Sagan is equally persuasive on the point that the character

---

[17] Possible scenarios are examined in George H. Quester, *Nuclear First Strike: Consequences of a Broken Taboo* (Baltimore, Md.: Johns Hopkins University Press, 2006). On the concept of a nuclear taboo, see Nina Tannenwald, *The Nuclear Taboo: The United States and the Non-Use of Nuclear Weapons Since 1945* (Cambridge: Cambridge University Press, 2007), esp. pp. 327–360.

[18] As Colin Gray has noted, a small nuclear war is an oxymoron. See Gray, *The Second Nuclear Age*, pp. 93–97.

[19] I gratefully acknowledge Gregory Treverton for this felicitous phrase. He bears no responsibility for its use here.

[20] Kenneth N. Waltz, "More May Be Better," Ch. 1 in Scott D. Sagan and Kenneth N. Waltz, *The Spread of Nuclear Weapons: A Debate* (New York: W.W. Norton, 1995), pp. 1–45.

of regimes and domestic policy making processes, as well as organizational aspects of military decision making, count for a great deal in nuclear crisis management.[21] The post-Cold War world may provide a plurality of regimes and political cultures that challenge the requirements for successful nuclear crisis management, especially the need for transitive expectations, clear communications and shared understandings about nuclear danger.

Defining minimum deterrence for a plurality of worlds poses a potentially open ended research agenda. The present international system, or possible iterations of it during the first quarter of the twenty-first century, offers a sufficient number of uncertainties and unknowns to challenge theorists and planners. What might minimum nuclear deterrence mean in the present and near term, given the inexorable weight of precedent on policy makers and on their available options? How viable might any minimum deterrence regime be, even if agreed to by the leading nuclear weapons states or all of them? The following discussion attempts to clarify these issues with conceptual boundaries and empirical referents.

Defining and measuring minimum deterrence

Defining minimum deterrence

The meaning of "minimum" deterrence is not necessarily obvious without having addressed the question of "compared to what?" Nuclear strategists would probably agree that minimum deterrence lies somewhere between assured destruction, as emphasized during Cold War discussions about nuclear strategy, and nuclear abolition. Exactly where is more debatable. At least four kinds of variables are in play in classifying nuclear strategies: (1), the political and military objectives for which forces are tasked; (2), the specifics of nuclear targeting plans, related to retaliatory objectives but not necessarily reflecting the actual intent of policy makers; (3), the numbers of weapons and launchers deployed and their assumed rates of survivability against first or later strikes; and, (4), the command-control systems and operational protocols of the state's nuclear forces, including their dependency on high states of alert or prompt launch for survivability. During the high Cold War, this might have lead to a spectrum of possible nuclear deterrent strategies as summarized in Table 8.1.

---

[21] Scott D. Sagan, "More Will Be Worse," Ch. 2 in Sagan and Waltz, *The Spread of Nuclear Weapons*, pp. 47–91. See also on these points: Richard Ned Lebow, *Nuclear Crisis Management: A Dangerous Illusion* (Ithaca, N.Y.: Cornell University Press, 1987), and Jervis, *The Meaning of the Nuclear Revolution*, Ch. 5.

Table 8.1 Attributes of generic nuclear deterrence strategies

	Counterforce-warfighting	Assured destruction	Minimum deterrence
Objectives and targeting	Victory or "prevailing" in a protracted conflict by imposing escalation dominance on the opponent at any phase	Inflicting retaliatory strikes sufficient to impose "unacceptable" damage on any attacker, including its remaining forces, C3, industry and population	Impose unacceptable damage to the attacker's society and civilian population and-or national infrastructure, although with forces less than those required for assured destruction
Numbers of weapons-launchers required	Numbers of survivable weapons capable of attacking or holding at risk military, C3, industry and population targets, if necessary through phases of a protracted war – may also require antimissile defenses for protecting population and-or forces-requires numbers of deployed warheads in the thousands, well above the threshold for assured destruction	Numbers of survivable weapons capable of attacking military, C3, industry and population targets and inflicting "unacceptable" damage – allows for flexible targeting but does not envision fighting a protracted nuclear war to a successful conclusion – requires numbers of deployed warheads in the thousands, fewer than required for counterforce-warfighting strategies	Numbers of survivable weapons sufficient to destroy major infrastructure and the sinews a modern national economy, while not necessarily emphasizing the destruction of urban-industrial areas, but also not necessarily guaranteeing "city avoidance" – requires numbers of deployed warheads in the hundreds

Table 8.1 (continued) Attributes of generic nuclear deterrence strategies

	Counterforce-warfighting	Assured destruction	Minimum deterrence
Command-control and alert-launch protocols	Political and military C3 must be not only survivable against initial attacks but enduring through various phases of a protracted conflict – some proportion of the force will be on hair trigger alert even in peacetime	Political and military C3 must be survivable for second strike retaliation and for post-attack negotiation for war termination – no forces on high alert required in peacetime but not precluded either	Political and military C3 must be survivable for second strike retaliation – no forces on high alert in peacetime

Source: Author. See also: Robert Jervis, *The Meaning of the Nuclear Revolution: Statecraft and the Prospect of Armageddon* (Ithaca, N.Y.: Cornell University Press, 1989), pp. 74–106; Scott D. Sagan, *Moving Targets: Nuclear Strategy and National Security* (Princeton, N.J.: Princeton University Press, 1989, esp. pp. 58–97; Desmond Ball, "The Development of the SIOP, 1960-1983," Ch. 3 in Desmond Ball and Jeffrey Richelson, eds., *Strategic Nuclear Targeting* (Ithaca, N.Y.: Cornell University Press, 1986), pp. 57–83; Jervis, *The Illogic of American Nuclear Strategy* (Ithaca, N.Y.: Cornell University Press, 1984), esp. Ch. 3–4; Ball, "U.S. Strategic Forces: How Would They Be Used?," in Steven E. Miller, ed., *Strategy and Nuclear Deterrence* (Princeton, N.J.: Princeton University Press, 1984), pp. 215–244.

The preceding table cannot capture all the nuances or possible variations within, and among, these three kinds of strategies. In addition, states' declaratory strategies are not always consistent with their operational policies.[22] But the table illustrates some of the qualitative and quantitative points of similarity and difference among these kinds of generic nuclear strategies.

For present purposes, minimum deterrence in today's world implies that U.S. and Russian arsenals would be limited to a maximum number of 1,000 operationally deployed strategic nuclear weapons, or fewer if possible. "Fewer if possible" means that for Washington and Moscow to go below 1,000 deployed weapons on transoceanic or intercontinental launchers, other

[22] On this point, see especially Desmond Ball, "The Development of the SIOP, 1960–1983," Ch. 3 in Desmond Ball and Jeffrey Richelson, eds., *Strategic Nuclear Targeting* (Ithaca, N.Y.: Cornell University Press, 1986), pp. 57–83; and Ball, "U.S. Strategic Forces: How Would They Be Used?," in Steven E. Miller, ed., *Strategy and Nuclear Deterrence* (Princeton, N.J.: Princeton University Press, 1984), pp. 215–244.

acknowledged nuclear weapons states would have to commit themselves to proportional reductions and-or limitations. Substrategic nuclear weapons, including tactical or operational weapons that are deployed on land, at sea or air delivered, have both political and military-operational contexts requiring separate discussion. There is certainly the possibility that, in any multilateral, constrained nuclear proliferation regime, some weapons of medium or intermediate range might have to be included as "strategic" based on their potential effects against likely regional adversaries.

Measuring minimum deterrence

The charts that follow permit us to examine the deterrence stability of two minimum deterrence regimes.[23] In the first case, U.S. and Russian strategic nuclear forces are limited to a maximum number of 1,000 operationally deployed weapons for each state. In the second case, a lower limit of 500 operationally deployed weapons is imposed on each.[24] For these larger and smaller minimum deterrent forces, we calculate their expected numbers of second strike surviving and retaliating warheads under four operational conditions of alertness and launch protocols: (1), generated alert, and launch on warning; (2), generated alert, riding out the attack and retaliating; (3), day to day alert, and launch on warning; and, (4), day to day alert, and riding out the attack. One might anticipate that, in general, the numbers of surviving and retaliating warheads would diminish as we proceed from option 1 through 4 above, but that progression is not necessarily automatic, depending on the specific circumstances of attack and response.

The results of this analysis appear in Figures 1 and 2, immediately below. Figure 1 summarizes the numbers of second strike surviving and retaliating warheads for the United States and for Russia, under each of the operational conditions listed above, for the case of 1,000 maximum deployed weapons for each country. Figure 2 provides information equivalent to that summarized in Figure 1, but for the more restrictive case in which maximum deployments are capped at 500 weapons for each.

---

[23] Grateful acknowledgment is made to Dr. James J. Tritten for the use of a model originally developed by him and modified by the author. He has no responsibility for its use here, nor for any arguments or conclusions in this study.

[24] Forces are hypothetical structures, although not unrealistic ones, used for analytical purposes, not predictions of actual deployment decisions. For pertinent estimates of U.S. and Russian future deployments, see Podvig, "New START treaty in numbers;" Podvig, "Russia's new arms development," *Bulletin of the Atomic Scientists*, January 16, 2009, http://thebulletin.org/web-edition/columnists/pavel-podvig/russias-new-arms-development; and U.S. Department of Defense, *Nuclear Posture Review Report* (Washington, D.C.: U.S. Department of Defense, April 2010).

The results summarized in Figures 1 and 2 show that either the 1,000-max minimum deterrence regime or the 500-max alternative provides for sufficient numbers of second strike surviving and retaliating warheads to guarantee unacceptable retaliation under each of the four operational conditions. Even in the worst case for defenders, with forces attacked while they are on "day to day alert, riding out the attack," Russian and American forces provide for several hundred retaliating weapons under a deployment limit of 1,000 and for more than one hundred retaliating weapons under a lower limit of 500 deployed warheads. However, under the conditions of any political crisis in which the United States and Russia were actually considering the use of nuclear weapons, both states' forces would doubtless be raised to higher alert levels and-or poised for prompt instead of delayed launch. In the canonical case often used for analysis (but not necessarily reflecting the likelihood of actual operations), either Russia or the United States, under conditions of "generated alert, riding out the attack," could provide for some 360 retaliating weapons (under a deployment limit of 1,000 warheads) or for about 190 retaliating warheads (under a lower deployment limit of 500 warheads).

These figures are hypothetical, although not implausible, outcomes for nuclear force exchanges under each of the two regimes. However, proposals to reduce U.S. or Russian forces to these post-New START levels may fail in politics despite the claims of analysts. One of the obvious speed bumps for

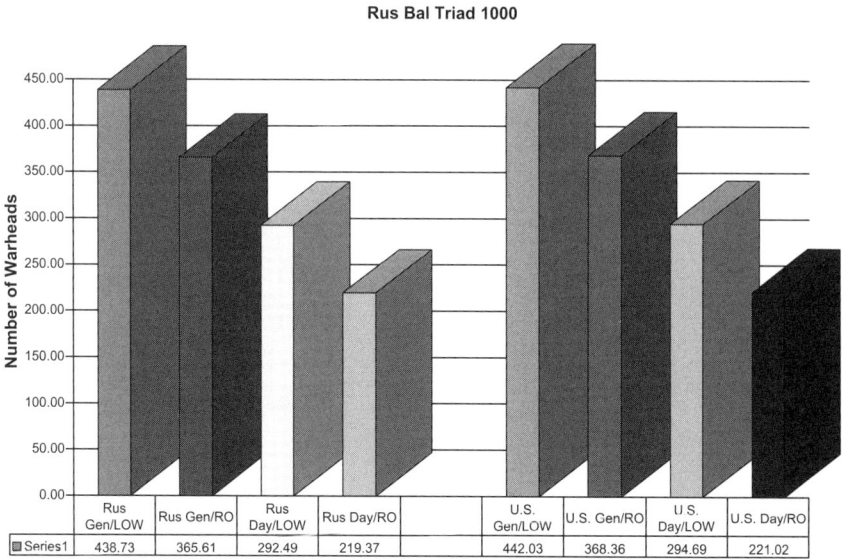

Rus Bal Triad 1000

	Rus Gen/LOW	Rus Gen/RO	Rus Day/LOW	Rus Day/RO		U.S. Gen/LOW	U.S. Gen/RO	U.S. Day/LOW	U.S. Day/RO
Series1	438.73	365.61	292.49	219.37		442.03	368.36	294.69	221.02

Figure 1 U.S.-Russia Surviving and Retaliating Warheads 1,000 Deployment Limit

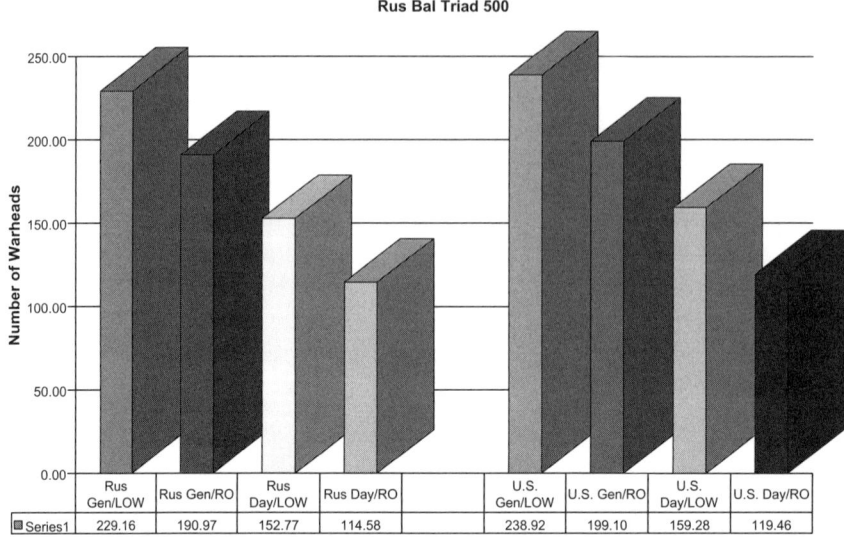

Rus Bal Triad 500								
	Rus Gen/LOW	Rus Gen/RO	Rus Day/LOW	Rus Day/RO		U.S. Gen/LOW	U.S. Gen/RO	U.S. Day/LOW	U.S. Day/RO
Series1	229.16	190.97	152.77	114.58		238.92	199.10	159.28	119.46

Figure 2   U.S.-Russia Surviving and Retaliating Warheads 500 Deployment Limit

Russia is the U.S. revised plan to deploy phased, adaptive missile defenses in Europe.[25] Russian leaders have insisted that they must be involved in U.S. and NATO missile defense planning, deployments and operations. During the NATO–Russia summit in Lisbon in November, 2010, Russian President Dmitry Medvedev agreed to future talks with NATO about joint missile defense deployments. In February, 2011 Medvedev appointed Russian ambassador to NATO Dmitri Rogozin as the special presidential envoy for missile defense, adding to the presumed diplomatic status of the issue.[26] On the other hand, both Medvedev and Russian Prime Minister Vladimir Putin warned in November, 2010 that any U.S.–NATO European missile defense plan that excluded Russia could lead to a nuclear arms race, including new deployments by Russia of offensive nuclear weapons and "strike forces."[27]

---

[25] Associated Press, "US launches new missile defense program for Europe," March 2, 2011, http://www.google.com/hostednews/ap/article/ALeqM5hulwA3apIb2YtB7eB; "U.S. Says Radar Ship Deployment Part of Missile-Defense Shield," *Radio Free Europe/Radio Liberty*, March 2, 2011, http//www.rferl.org/articleprintview/2325314.html.

[26] Vladimir Frolov, moderator, "Russia Profile Weekly Experts Panel: Russia Raising the Stakes on Missile Defense," *Russia Profile*, February 25, 2011, in *Johnson's Russia List* 2011 – #36, February 28, 2011, davidjohnson@starpower.net.

[27] "Warning of New Arms Race, Medvedev Calls for Cooperation with West On Missile Shield," *Radio Free Europe/Radio Liberty*, http://www.rferl.org/content/russia_medvedev_parliament/2234566.html.

Russia's objections to U.S. missile defenses deployed in Europe under NATO's aegis have more to do with politics than with the logic of nuclear deterrence.[28] The inferiority of Russia's conventional forces compared to those of NATO makes Russia more reliant on its nuclear forces for missions other than deterrence of a U.S. or NATO nuclear first strike. Russia's military doctrine allows for the first use of nuclear weapons by Russia in a conventional war that includes attacks near Russia's periphery or into Russia's state territory with the potential to jeopardize Russia's vital interests and sovereignty.[29] Russia in particular fears NATO's capabilities for conventional deep strike missions and the alliance's relative superiority in information based technologies for conventional warfare. However improbable or illogical these Russian concerns might seem from a U.S. or NATO perspective, Russia's sense of conventional military inferiority invites its military planners to fill the gap with Russia's nonstrategic nuclear weapons for deterrence and escalation control.

Politics as well as military art also dictates that Russia holds fast to its image of strategic nuclear parity with the United States. This perception, of Russia and the United States sharing a singularity in strategic nuclear capabilities compared to other powers, carries political overbite for Russian negotiators in various international forums and provides for Russia a toehold on great power military status. Russia's sensitivities about U.S. missile defenses are as much about this perception of Russian–American strategic nuclear equivalence regardless of military-technical realities. Thus fears expressed by Russia's politicians and military divas about a creeping U.S. nuclear first strike capability are not based on realistic perceptions of American intentions.[30] Instead, these sentiments perform two functions in Russian domestic politics. First, the Russian general staff can continue to use NATO and the United States as bell ringers in threat assessments. Second, NATO-centric threat assessments help

---

[28] Yuri Solomonov, expert missile designer at the Moscow Institute of Thermal Technology, told a press conference in March, 2011 that the proposed U.S. missile defense plan for Europe presented no threat to Russian strategic nuclear forces. He added that European missile defense agitation "is created by politicians on one side and the other for gaining certain concessions and resolving totally unrelated problems with package agreements." See "U.S. missile defense in Europe does not threaten Russia – Solomonov," *Interfax*, March 17, 2011, in *Johnson's Russia List 2011* – #50, March 18, 2011, davidjohnson@starpower.net.

[29] "The Military Doctrine of the Russian Federation," text, www.Kremlin.ru February 5, 2010, in *Johnson's Russia List* 2010 – #35, February 19, 2010, davidjohnson@starpower.net.

[30] U.S. modernization might create a situation of Russian nuclear inferiority or jeopardy regardless of U.S. intentions, according to some analysts. See Keir A. Lieber and Daryl G. Press, "The Rise of U.S. Nuclear Primacy," *Foreign Affairs*, March/April 2006, http://www.foreignaffairs.org/20060301faessay85204/keir-a-lieber-daryl-g-press/html. Rejoinders to Lieber and Press include Peter C.W. Flory, Keith Payne, Pavel Podvig, and Alexei Arbatov, "Nuclear Exchange: Does Washington Really Have (or Want) Nuclear Primacy?" *Foreign Affairs*, September/October 2006, http://www.foreignaffairs.com/print/61931.

to forestall the transition from a mass mobilization army based on conscripts to a professional army, the latter structured around brigades manned with voluntary contract soldiers and trained for rapid deployment into hybrid wars with conventional and-or unconventional features.

Politics excepted, are Russian concerns about future NATO missile defense capabilities entirely self serving? By the last phase of Obama's European missile defense plan in 2020, U.S. BMD technology will presumably have improved over present models. Fourth generation SM-3 interceptors and supporting C4ISR could conceivably have some intercept capabilities against intercontinental missiles launched from Russia or elsewhere, especially if the missile defense launchers were widely deployed across terrestrial and maritime space. On the other hand, whether the Obama plan provides "game changing" missile defenses also depends upon Russia's fulfillment of its offensive missile modernization plans, including possible countermeasures against defenses. An additional complication is that futuristic antimissile defenses will have some commonality with technologies also contributory to air defenses against bomber attack. Further uncertainty exists in the politics of NATO decision making, with respect to which member states will host missile interceptors or other components of the regional missile defense system – with the possibility that those hosts will feel Russian pressure or even threats of targeting by Russian nuclear forces.

With the preceding considerations in mind, analysis of the possible impact of missile defenses on minimum deterrence stability is now in order. In Figures 3 and 4, below, we summarize the results of nuclear exchanges in which U.S and Russian second strike retaliatory forces are opposed by antimissile and antiair defenses (combined) with variable intercept and destruction capabilities. In this example, Phase I defenses successfully intercept (destroying or deflecting attacking re-entry vehicles from intended targets) 20 per cent of the retaliators – or, from the opposite standpoint, permit a leakage of 80 per cent. Phase II defenses intercept 40 per cent and permit a leakage of 60 per cent. Phase III defenses intercept 60 per cent, and Phase IV defenses, some 80 per cent. Each set of defenses is played against a U.S. and Russian second strike retaliation based on forces on generated alert and riding out the attack before retaliating. The results of this analysis are summarized for U.S. and Russian forces, under a peacetime deployment limit of 1,000 warheads, in Figure 3. Figure 4 provides similar information about outcomes for Russian and U.S. forces, within a lower peacetime deployment ceiling of 500 warheads.

The results summarized in Figures 3 and 4, above, show that either the larger or smaller minimum deterrent force would provide for sufficient numbers of "disasters beyond history" in second strike retaliation, even if opposed by highly competent antimissile and antiair defenses by today's standards. The

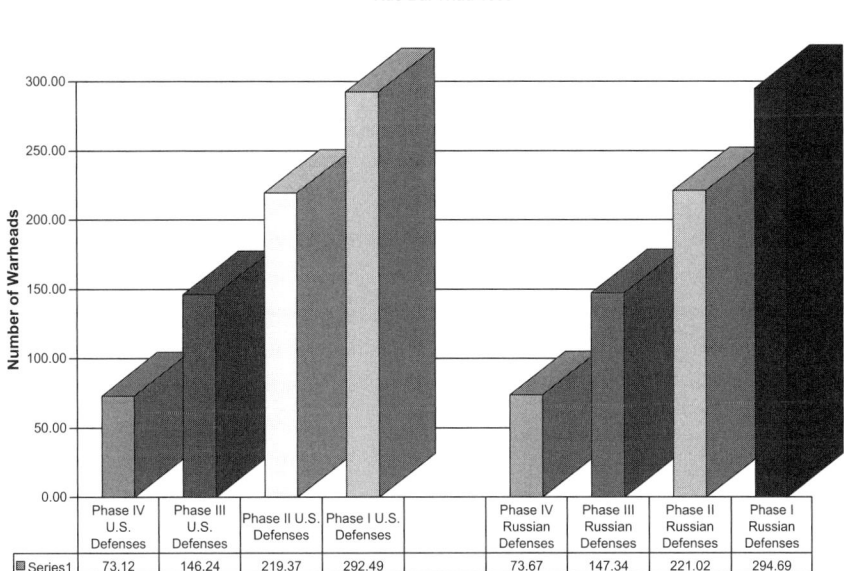

Rus Bal Triad 1000

	Phase IV U.S. Defenses	Phase III U.S. Defenses	Phase II U.S. Defenses	Phase I U.S. Defenses		Phase IV Russian Defenses	Phase III Russian Defenses	Phase II Russian Defenses	Phase I Russian Defenses
Series1	73.12	146.24	219.37	292.49		73.67	147.34	221.02	294.69

Figure 3 U.S.-Russia Surviving and Retaliating Warheads vs. Defenses 1,000 Deployment Limit

surviving, retaliating and defense-penetrating warheads for both states would not provide for much flexibility in targeting, especially under the assumption of a 500 warhead peacetime deployment limit. Fighting counterforce wars or protracted nuclear conflicts would be neither feasible nor, frankly, desirable. Nuclear "withholds" for postattack escalation control might be included in planners' prewar designs, but minimum deterrents would regard such forces as assets only for negotiation and war termination – not for continued fighting.

A case can be made that deploying U.S.–NATO or Russian defenses is necessary in order to help deter or defeat attacks from nuclear hostiles such as Iran or North Korea. Defenses can provide insurance against the consequences of light attacks although those same technologies could not preclude an American or Russian second strike, thereby leaving intact a mutual deterrence relationship between Washington and Moscow. This logic is fine as far as it goes, but the more important political question is whether nuclear "deterrence" or "cooperative security" is the appropriate paradigm for future U.S.–Russian negotiations – with, or without, missile defenses. It would be ironic if NATO–Russian collaboration on European theater missile defense accelerated the Obama "reset" and further disestablished nuclear "deterrence" as the anchor baby of U.S.–Russian military relations.

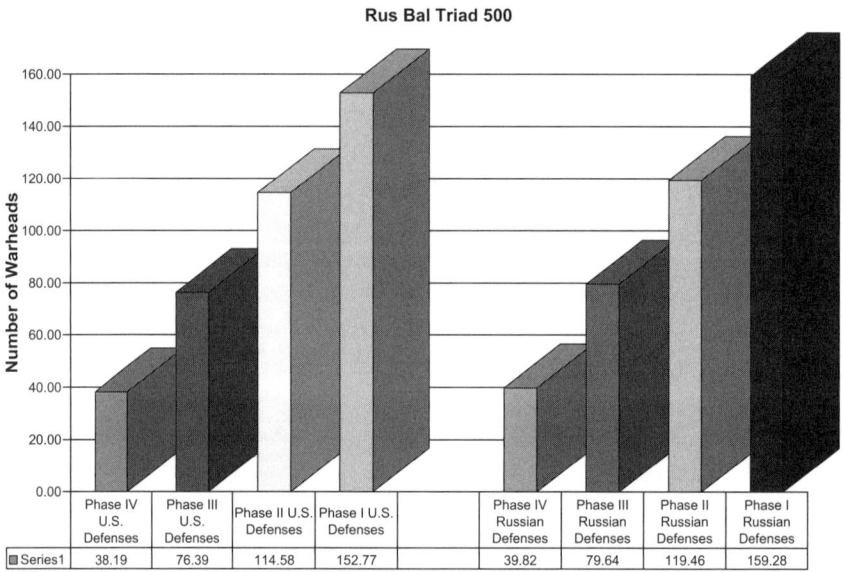

Figure 4 U.S.-Russia Surviving and Retaliating Warheads vs. Defenses 500 Deployment Limit

Conclusions

As between the United States and Russia, a minimum deterrence regime with a maximum number of 1,000 or 500 deployed long range weapons could certainly provide for adequate numbers of surviving and retaliating weapons to ensure deterrence and crisis stability. Although some advocates of minimum deterrence might prefer the lower limit to the higher one, accurate appraisal of Russian and U.S. domestic politics (including the uncertainties of looming elections in both countries) suggests optimism tempered by realism. In addition, going significantly below 1,000 operationally deployed strategic nuclear weapons for either Moscow or Washington will require the cooperative restraint of other nuclear weapons states. As political relations between the two states improve, the probability increases for an agreed minimum deterrence standard. Nor is it precluded that the emergence of a genuine "security community" across the breadth of Europe, including NATO and Russia, would make nuclear deterrence within that community unnecessary or undesirable.[31] However, Russia will not be satisfied with nuclear parity per se: its nuclear ramparts must cover its basic and extended deterrence wagers until its conventional force modernization moves beyond the aspirational level into the operational realm.[32]

Agreement on a minimum deterrent standard for a fixed number of accepted nuclear weapons states, while drawing a firm line against others joining that club, appears a controversial choice until the alternatives are considered. The most probable alternative is to rely on traditional patterns of power politics, supplemented by a nonproliferation regime of gradually diminishing status and by unilateral guarantees of extended deterrence, for the avoidance of both nuclear proliferation and nuclear war.

---

[31] Some have argued that the U.S. and Russia are ready for this step now. See Samuel Charap and Mikhail Troitskiy, "Time to Put an End to MAD," *The Moscow Times*, January 18, 2011 http://www.themoscowtimes.com/opinion/article/time-to-put-an-end-to-mad/429055.html.

[32] Olga Oliker, Keith Crane, Lowell H. Schwartz and Catherine Yusupov, *Russian Foreign Policy: Sources and Implications* (Santa Monica, Calif.: RAND, 2009), pp. 139–174, esp. p. 164 on the Russian need for nuclear weapons to cover conventional deterrent missions.

Conclusions

I. Reviewing the Findings

Everything old is new again – and vice versa. Nuclear weapons, regarded by some as atavistic survivors in a postindustrial age of precision warfare and reduced collateral damage, have nevertheless attained a niche status in policy making and military-strategic planning in the present century. The nuclear weapons of mass destruction that some credited with helping to stabilize the Cold War are now respected and feared for their deterring or destructive power in the hands of rogue states, terrorists or other nuclear armed combatants with angry agendas. Nuclear deterrence, and therefore the possibility of nuclear war, coexists uneasily with conventional warfare based on information principles and advanced cyber technologies. The top down command-control required for the management of nuclear forces, so that they are responsive to lawful commands but protected against accidental or unauthorized use, contrasts markedly with the network-centric principles of distributed command-control in the information age. In this section, we summarize the "lessons learned" from the preceding chapters and their implications for further study and policy making. The main arguments of each chapter, in summary form, are as follows.

Chapter 1

Nuclear weapons will not exist in a vacuum of politics or policy. Some context for the relationship among institutions, norms, and values, and behaviors relevant to nuclear weapons and nuclear deterrence, will be provided by one or more constructs that we call alternative nuclear regimes. Descriptions of these regimes include assumptions about the relative sizes of states' nuclear arsenals and the ability of the international community to contain the spread of nuclear weapons going forward. Whether any of these regimes holds pride of place in the first decades of the twenty-first century depends as much, or more, on politics, compared to technology. Missile defenses and other advanced conventional weapons may create additional possibilities for the attainment of strategic purposes by means of "deterrence by denial" in addition to nuclear reliant deterrence by threat of unacceptable retaliation and punishment. But advanced conventional weapons will not repeal the nuclear revolution that privileges the assured vulnerability of populations to nuclear missile attacks, unless, and until, entirely new generations of antinuclear countermeasures can be devised.

Chapter 2

The information age has ushered into military parlance and practice a keen interest in cyberwar, including offensive attacks by a state against another state's military or civilian targets. Cyberwar is certainly included in the tool kit of postindustrial advanced militaries, including the U.S. armed forces. The U.S. Department of Defense established a Cyber Command in 2009, and evidence is abundant of American, Russian, Chinese, and other great power interest in offensive and defensive cyberwar. The topic of cyberdeterrence, somewhat more abstract and undefined compared to cyberwar, is nevertheless also a candidate master concept for militaries with information-based weapons and C4ISR (command, control, communications, computers, intelligence, surveillance and reconnaissance) systems. Some cyber attacks support forces in combat (operational cyberwar) and others can aspire to strategic effects on their own terms against enemy information systems or information-reliant infrastructure, including electric power grids, transportation and finance.

However, the use of cyber warfare by nuclear armed disputants in a crisis may be dangerous and counterproductive to otherwise successful nuclear crisis management. Nuclear crisis management anticipates both competitive and cooperative behavior on the part of disputants. Transparency of decision making processes, clear and uninterrupted communications, accurate information about opponents' true intentions and capabilities, and an ability to avoid self imposed or other arbitrary deadlines for making a decision, are all important supports for resolving a crisis peacefully. A nuclear armed state faced with a sudden burst of holes in its vital warning and response systems might, for example, press the preemption button instead of waiting to ride out the attack and then retaliate. Nuclear weapons states involved in a serious political crisis would want to ascertain the alert and launch readiness status of one another's forces, and the preparedness of command-control systems for war, in order to avoid either (1), a mistaken launch of nuclear retaliatory forces, under the assumption that the other side had already attacked or ordered an attack; or, (2), a mistaken decision not to retaliate that maximized the damage caused by the enemy attack. In addition, the very anticipation of information warfare against networks and command-control systems by another state might lead to preemptive cyber "retaliation" spilling over into kinetic attacks, either conventional or nuclear. Cyberwar and nuclear weapons may not play well together, although cyberdeterrence might operate with some prudence, as between cyber-dependent states locked into a conflict spiral.

Chapter 3

Geography is part of the environment in which any decision related to international security and nuclear weapons takes place. Nuclear weapons are located somewhere, aimed at something, programmed to follow certain trajectories from point of the origin to the destination, and so forth. Nuclear weapons (thus far) are deployed by states, and states have territorial, that is, geographical, definitions as part of their DNA. Since geography is inescapable from decisions about nuclear force development, deployment and use, nuclear strategy is partly geostrategy. Nuclear geostrategy is complicated by the deployment of nuclear weapons and launch platforms in diverse basing modes: on land; at sea; and airborne. Diverse basing modes require complex arrangements for stable and secure nuclear command and control, as against the possibility of accidental or inadvertent war or, the opposite, a failure on the part of nuclear forces to respond in a timely manner to authorized commands. U.S. and former Soviet global nuclear reach set them apart from other nuclear armed combatants during the Cold War, and Russia still claims approximate nuclear-strategic parity with the United States on the basis of its numbers of intercontinental range launch systems. However, future international stability is endangered less by the prospect of global nuclear war than it is by regional conflicts between nuclear armed disputants – especially if nuclear weapons spread continues in Asia.

Chapter 4

Nuclear abolition is an idea whose time has come or so it seems. U.S. President Barack Obama's declarations in favor of eventual nuclear abolition have been followed by a cautious approach to international security issues, including the successful negotiation of a New START agreement making important, albeit modest, reductions below previously agreed levels of deployed weapons and launchers. The obstacles to nuclear abolition are less ideological than they are practical. In practical terms, what does nuclear "abolition" require of states? Nuclear abolition could mean no deployed weapons, no assembled weapons, or no weapons grade materials that are not under foolproof international control. If the first objective is the goal, then no nuclear weapons will be deployed on launchers, although they may be assembled and stored. This condition might be an improvement over the present situation – if it were stable. But states' distrust might soon erupt into a competition to accumulate larger quantities of assembled and stored weapons – a clandestine nuclear arms race, with the potential for eventual breakout into an overt competition again. It seems clear that new politics are required for movement from the present condition to nuclear abolition. Either the leading nuclear weapons states would have to drive other states into a consensus for nuclear reductions to zero, or some international authority would have to supervene over states' powers in this regard. Even the realization of true nuclear abolition would not be uncontroversial – critics would contend that it favored the leading powers in advanced technology, conventional warfare, especially the United States

Chapter 5

The New START agreement signed by Presidents Barack Obama and Dmitri Medvedev in April, 2010, and ratified by the U.S. Senate in December, 2010, represented an important first step in the "reset" of U.S.–Russian relations and possible precursor of further Russian–American cooperation on nuclear arms reductions and nonproliferation. The significance of New START lies, less in the numbers of warheads, launchers and so forth, and more in the reassurances that the numbers provide about the two states' intentions. New START may be the first of several rounds of nuclear arms reductions as between Moscow and Washington. Some proposals are already being floated for reductions below the New START agreed limits: ceilings of 1,550 deployed warheads and 700 deployed intercontinental launchers. However, reductions below New START may involve greater domestic opposition, especially in Moscow. Russia wanted to include language in New START that explicitly tied offensive arms reductions to limits on U.S. missile defenses, but the Obama administration refused, and Russia settled for a declaratory statement that offenses and defenses are linked in principle. Some analysis shows that the United States and Russia could maintain stable deterrence with forces reduced to a maximum of 1,000 deployed long range warheads for each state, or even fewer. But U.S. and Russian reductions below 1,000 weapons would require the cooperation of other nuclear weapons states in making proportional cuts in their deployments – something they might be loath to do. Moreover, Russia maintains a healthy skepticism about U.S. missile defense modernization plans, and the Obama reset of Bush's original missile defense plans for Europe has not necessarily mollified Kremlin hawks on this point. Reductions in offensive weapons below New START levels will probably require more explicit consideration of U.S.–Russian cooperation on missile defenses, perhaps including the inclusion of Russian radars or other components in a European-wide missile defense system.

Chapter 6

North Korea fulfills Churchill's description of the Soviet Union as a riddle wrapped in a mystery inside an enigma. North Korea's membership in the ranks of nuclear weapons states is opposed by the United States and other important actors in Asia, including Russia, China, Japan, and South Korea. All five have been engaged in off-and-on, multilateral talks with the DPRK, with the objective of eventual elimination of North Korea's nuclear weapons, its weapons grade materials and its nuclear infrastructure. There is little appetite among the powers for military action against North Korea, which would be very costly for South Korea (even in "victory"), and against a regime that despoils its own population and values only its own survival in power. The U.S. and other negotiators have tried a mixture of carrots and sticks against North Korea with regard to nuclear disarmament, but every apparent step forward runs into eventual stonewalling or abrupt changes in the terms of trade. The United States and other interlocutors with North Korea might succeed by enlarging the scope of the discussions. The U.S., China, South Korea and North Korea could agree an official termination for the Korean War, including a peace treaty that recognizes the permanence of the regime in Pyongyang and offers additional economic and political incentives in return for denuclearization. Short of that, interim strategies for managing the North Korean situation may require scenario planning based on multiple, hypothetical "North Koreas" and unexpected events – including a sudden collapse of the Kim regime or its fractioning into civil war.

Chapter 7

Although discussion in earlier chapters noted some of the dangers attendant to nuclear weapons spread in Asia, Chapter 7 reminds us that complacency about the possibility of nuclear danger in Europe is ill advised. Post-Soviet Russia is not in an officially hostile relationship with the United States or NATO, and NATO has expanded to twenty-eight member states across the breadth and depth of Central and Eastern Europe. However, debates still appear in prominent political and military circles about the role of nuclear weapons deployed in Europe or capable of being brought to bear in a European conflict. The United States deploys several hundred so-called tactical or substrategic nuclear weapons on six bases in five NATO countries: Belgium, Germany, Italy, the Netherlands, and Turkey. Russia is estimated to have deployed several thousand substrategic nuclear weapons, many located in its western military districts. The United States, NATO and Russia have refused to announce a declaratory policy of "no first use" of nuclear weapons, although the Obama administration Nuclear Policy Review of spring, 2010 did argue for an eventual, ideal posture of "deterrence only" with respect to nuclear weapons. On account of its inferiority compared to the United States and NATO in advanced technology conventional weapons, Russia relies on its nuclear weapons to cover a greater spectrum of deterrence than did the former Soviet Union. Accordingly, Russia's Military Doctrine of 2010 acknowledges that nuclear first use is an option for situations in which Russia's survival as a state is at issue. In addition, earlier editions of Russian doctrine and some statements from prominent officials have suggested that Russia might resort to nuclear weapons as a means of strategic de-escalation to avoid losing a conventional war on, or near, Russian territory and threatening to Russia's vital interests. If, against current political odds, Russia and NATO somehow managed to become embroiled in a conventional war, the issues of nuclear "first use" and "first strike" would become intellectually and operationally tangled. The first use of a short range or tactical nuclear weapon in Europe would reformulate political and military expectations on the spot and provide a stark reminder that nuclear "escalation control" is not even a practiced art, let alone, an applied science.

Chapter 8

In Chapter 8, we ask whether the United States and Russia could advance beyond their accomplishments in New START by rethinking their underlying nuclear deterrence concepts. In particular, Russia and the United States could move toward a regime of "minimum deterrence" based on maximum numbers of operationally deployed strategic nuclear weapons in the hundreds instead of thousands. Improvements in the political climate between the United States and Russia since the end of the George W. Bush and Vladimir Putin presidencies may continue into Obama's second term and after Putin's return to the presidency of Russia. If so, minimum deterrence might emerge as a discussable alternative between Washington and Moscow, and as the basis for a multilateral, constrained nuclear proliferation regime including eight nuclear weapons states (excluding North Korea). Minimum deterrence as the basis for a multilateral nuclear "concert" might be established in two phases: first, among the P-5 permanent members of the UN Security Council; and, second, among the remaining three acknowledged nuclear weapons states, excluding North Korea. Our analysis shows that such a regime is not impossible, but neither is it guaranteed to have sufficiently favorable political winds.

II. History has a sense of humor – including macabre

The novelty of "the bomb," or at least, of thinking about nuclear weapons compared to others, was summarily expressed in the lingua franca of post-World War II military strategizing. The term "deterrence," borrowed from psychology, became a focal concept, nearly a paradigm shift, in how the threat or use of nuclear force was to be managed within the context of an international power struggle. With nuclear weapons in the background, the risk of "total war" in which countries threw the entirely of their armed might into the battle *a l'outrance*, would be unacceptably self destructive. Once the large and mature American and Soviet strategic (transoceanic or intercontinental range) nuclear forces were deployed in great numbers, on multiple launch systems, and in survivable modes, a staring contest in the manipulation of the threat of war replaced the actual use of weapons for the purpose of prevailing in combat. In addition, the vast superiority of Soviet and U.S. nuclear forces compared to others, and their respective bloc leadership in and outside of Europe, helped to create a closed loop of political and military stability that lasted until the end of the Cold War.

It was anticipated by some expert and lay audiences that the end of the Cold War and the demise of the Soviet Union would lead directly to the marginalization, or even the elimination, of nuclear weapons in the twenty-first century. For those who so hoped, disillusionment has been unavoidable, despite understandable impatience. Since nuclear weapons were so dominant in military thinking in the years between 1945 and 1991, it seemed to follow logically that they would be less significant in the New World Order. To some extent, this was so. A "revolution in military affairs" that privileged information-based warfare, including systems for conventional deep strike and smart C4ISTAR systems (command, control, communications, computers, intelligence, surveillance, targeting and reconnaissance), transformed the

United States into a singularly global military power. However, the United States did not give up its nuclear weapons, and continued to modernize its strategic nuclear forces – as did Russia, although under economic difficulties of special duress during the 1990s. In addition, India, Pakistan and North Korea declared themselves into the club of nuclear weapons states in the first two decades after the end of the Cold War, and other states, including Iran, hovered on the threshold. Fears of a nuclear arms race in the Middle East rivaling the already underway nuclear competition in Asia, were realistic and rising as U.S. President Barack Obama assumed office in 2009.

Obama had been preceded in the White House by George W. Bush, and the Bush-43 administration had offered its own solutions to the relationships among the four horsemen of nuclear danger. Bush regarded the problem of nuclear danger neither as one caused by technology per se nor, as his critics would argue, by the weakness of international nuclear control regimes. Instead, for the Bush-43 administration, the link between nuclear weapons and global insecurity was based on the character of the regimes which had acquired, or might acquire, nuclear or other weapons of mass destruction. Bush was not opposed to international sanctions and other measures short of war in constraining WMD proliferation. In addition, however, the United States asserted the prerogative to impose regime change on states that harbored terrorists, acquired WMD with the possibility of passing it along to terrorists, or threatened to use WMD for attacks on the United States or its allies. This prerogative as asserted by Bush-43 included the option for preemptive or preventive war, amid great controversy with allies and domestic critics. Regime change was imposed on Iraq in 2003 (despite objections from permanent members of the UN Security Council and some major European allies of the United States), based on the above rationale: that the character of nuclear danger lay not in the "what" of nuclear technology or the "how" of nuclear regulatory regimes, but in the "who" of nuclear ownership and the owners' presumed intentions.

The George W. Bush administration also argued that the end of the Cold War and the demise of the Soviet Union made the Cold War approach to nuclear arms control obsolete. There were two aspects to the Bush departures from precedent in this regard. First, Bush-43 announced U.S. intent to abrogate the ABM Treaty and to begin deployment in 2004 of components for an eventual global missile defense system. This system was not intended as a new version of the Reagan Strategic Defense Initiative, proposed in 1983 as a research and development program that would eventually provide a comprehensive defense for the continental United States against a massive Soviet nuclear attack. Instead, the Bush-43 missile defense plan had more continuity with the Bush-41 and Clinton programs for limited defenses against light attacks from "rogue" states or accidental launches. During Clinton's second term, the U.S. Congress passed legislation calling for ballistic missile defense (BMD) deployments as soon as workable technology was available, but Clinton demurred to act on this Republican-favored plan before leaving office.

In addition to withdrawing from the ABM Treaty and beginning deployment of missile defenses, the Bush-43 administration favored a more streamlined and less bureaucratic approach to U.S.–Russian nuclear arms control. Arguing that Cold War treaties were too encumbered by detailed protocols for verification and inspection, the Bush-43 administration agreed with Moscow in 2002 the Strategic Offensive Reductions Treaty (SORT). This agreement limited both states to a maximum of between 2,200 and 1,700 deployed nuclear warheads on missiles and bombers of intercontinental range. However, SORT received less than rave reviews from arms control experts, since it relied on verification procedures left over from the START I agreement signed in 1991 between the United States and the former Soviet Union. In addition, SORT had the odd aspect that its limits on warheads and launchers would go into effect only on the last day of the calendar year 2012, but with no further stipulation as to the lowering, or raising, of the required thresholds after that date.

The Barack Obama administration took office in January, 2009 and determined to change the Bush-43 approach to nuclear arms reductions, defenses and nonproliferation in several respects. First, the Obama administration revived the traditional approach to nuclear arms reductions, engaging with Russia in negotiations for a successor agreement to START I. In their July, 2009 summit in Moscow, Obama and Russian President Dmitri Medvedev agreed general guidelines for a post-START agreement that could reduce both states' numbers of deployed nuclear weapons to a range from 1,675 to 1,500 each. In addition, the two states sought a post-START verification regime that would bring forward the "best practices" of START I while doing away with some of its more anachronistic or superfluous restrictions. After considerable numbers of hard point negotiations and back and forth public diplomacy, Presidents Obama and Medvedev signed the New START agreement in April, 2010. The two states had to compromise on a number of matters in order to reach the end zone, including counting rules for warheads and launchers and Russian concerns about possible American "upload potential" based on stored, but not destroyed, nuclear weapons.

Second, the Obama administration revised the Bush-43 missile defense plan so that it was less politically objectionable to Russia. During 2007 then Russian President Vladimir Putin had declared his strenuous objections to U.S. plans for deployment of missile defense components in Poland in the Czech Republic. Putin interpreted the proposed Eastern European missile defenses as a threat to Russia's strategic nuclear deterrent and as a case of American and NATO political and military thrust into Russia's self declared privileged security space (states on its periphery and/or part of the former Soviet Union). Russia threatened to depart from both nuclear and conventional arms control agreements unless the United States reconsidered the European missile defense plan, especially the deployment of ground based interceptors in Poland. The Obama administration revised the Bush-proposed missile defense plan, revisiting its "third position" in Europe that Moscow found so objectionable. In September 2009, Obama proposed that the initial phase of European BMD deployments would consist of sea based SM-3 interceptors directed against the possibility of an Iranian or other Middle Eastern attack and not technically capable of compromising Russia's nuclear deterrent.

The Bush justification for the European based missile defenses had also been for the protection of U.S. and allied forces and other targets against a possible attack from Iran. However, the Obama administration sought additional "street cred" from its improved political relations with Russia and from the SM-3s' already proven record in service with United States Ticonderoga class cruisers and Arleigh Burke destroyers. Thus, according to U.S. defense officials, the new system would be in service sooner than the Bush-proposed system. The first phase of the Obama European BMD system would be composed entirely of sea based SM-3s (2011), although later stages could include the deployment of land based SM-3 missiles (by 2015). The possibility of enhancing SM-3 intercept capabilities from the short-to-intermediate range to include intercontinental missiles was not excluded for an estimated time frame of 2020. The Obama administration also held out to Russia the possibility of cooperation on missile defense technology, including shared early warning and networking between Russian and U.S.–NATO missile defense system components. Much would depend on the character of U.S.–Russian political relations, including further progress on nuclear arms reductions and nonproliferation.

The third aspect of the Obama "reset" with respect to nuclear issues and Russia was the problem of nonproliferation. In this area, as in nuclear arms reductions and in missile defenses, the U.S. and Russia had both cooperative and conflicting interests. In principle, both agreed that North Korea should be persuaded to reverse its decision for nuclear weaponization. Russia also contends that it shares the United States desire to prevent Iran from becoming a nuclear weapons state. But in practice, the two states have diverged in their diplomatic approaches to obtaining these objectives – especially with respect to Iran.

Russia has strong economic interests in Iran, including support for the construction of its nuclear reactor at Bushehr and apparent Iranian interest in purchasing advanced Russian air defense systems. On the other hand,

Russian territory lies nearer the launch points of Iranian missiles than does Western Europe or North America. Thus Russia prefers a mix of coercion and conciliation with Iran that is heavier on the latter, whereas the U.S. and the European contact group that has been negotiating with Iran about its nuclear program favor an approach involving more coercion. However, Iran remained officially defiant in the face of carrots and sticks, including three rounds of UN sanctions: denying its intent for eventual weaponization; refusing a UN-brokered plan for returning most of its low enriched uranium to Russia and France for reprocessing; and, third, announcing a planned expansion of its allegedly peaceful nuclear enrichment facilities.[1] Obama's plan for nonproliferation based on diplomatic "engagement" with Iran fell short of expectations. Thus, in June, 2010 the United States and eleven other members of the UN Security Council voted in favor of a fourth round of sanctions against Iran's nuclear program, targeting especially the military purchases, trade and financial transactions by Iran's Islamic Revolutionary Guards Corps (IRGC).[2]

The dustup with Iran over its nascent nuclear capabilities showed some of the interactions among the issues of nuclear arms reductions, missile defenses and nonproliferation. U.S. missile defenses deployed in Europe were justified by Bush and Obama on the basis of a future Iranian missile threat, including a possible nuclear one. Putin and his military advisors reacted to the Bush BMD plan by dismissing this U.S. rationale as a façade, behind which lay an American quest for nuclear superiority and the possible negation of Russia's deterrent. The empowerment of Iran as a regional threat, with or without nuclear weapons, had been expedited by the Bush-43 decision to impose regime change in Baghdad – removing Saddam Hussein as a source of regional containment against Iran. This, together with the U.S. takedown of the Taliban in Afghanistan in 2001, left Iran as the Shi'a superpower with bulging muscles capable of being flexed in either of two directions – westward, toward Iraq, or eastward, toward Afghanistan. Both the Obama and Bush-43 administrations, moreover, not only regarded Iran as a potential military threat, but also as a retro political regime standing in the way of further democratic enlargement in the Middle East.

Getting Russia's cooperation on Iranian denuclearization required getting past Russia's pique over missile defenses in Eastern Europe and vexation over the U.S. decision for regime change in Iraq. Without Russia's cooperation, successful diplomatic coercion of Iran short of military attack was improbable

---

[1] Iran Watch, "Iran's Nuclear Timetable," updated December 2, 2010 and regularly, http://www.iranwatch.org/ourpubs/articles/iranucleartimetable.html.

[2] Neil MacFarquhar, "U.N. Approves New Sanctions to Deter Iran," *New York Times*, June 9, 2010, http://www.nytimes.com/2010/06/10/world/middleeast/10sanctions.html

of success. The same held true, to an extent, for turning back the clock on North Korean nuclear proliferation. However, in the case of North Korea, China's immediate economic influence and national interests provided more direct leverage with Pyongyang than Russia had. North Korean nuclear proliferation, like that of Iran, was dangerous, not only on account of the character of the regime and its presumed intentions, but also because of the possibility of contagion effects in the region. A North Korean nuclear weapons state could export nuclear materials, technology and know-how to other states in Asia or the Middle East, or to terrorists. Even short of that, in reaction to nonreversible North Korean nuclear weapons status, South Korea, Japan or even Taiwan might reconsider their previous decisions to remain outside the nuclear club. In a similar fashion, Iran might provide nuclear materials, technology and expertise to other states or stateless movements in the Middle East and South Asia. In response to Iran's nuclear weapons status, Egypt, Saudi Arabia and Turkey might go nuclear, fearful of a rising Shi'a-majority Iran intent upon regional hegemony.

Missile defenses might be of interest to states in Asia or the Middle East as an alternative to having their own offensive nuclear weapons. Japan and the United States have cooperative agreements on missile defense as part of U.S. security guarantees to that country, and the United States and Israel have shared missile defense technologies and training. The U.S. reborn missile defense plan for Europe, as above, is based on technology that could be applied to defend allies outside of Europe. In either case, inside or outside of Europe, U.S.-provided or supported missile defenses could help to extend American deterrence, against nuclear blackmail of friends and allies by regional rogues or aspiring hegemons. Defenses would add the factor of denial of enemy objectives, in attacking with nuclear or other WMD-armed ballistic missiles, to the influence of deterrence by threat of retaliation. And, of course, defenses could provide insurance against accidental or inadvertent launches.

But antimissile defenses, unless and until based on different physical principles than hitherto, will be vulnerable to offensive countermeasures: including saturation, decoys and bypassing the defenses entirely with a more indirect approach (e.g. terrorism). Missile defenses will not repeal the nuclear revolution, nor will they slake the thirst of aggressive states that want nuclear weapons for political blackmail, for international prestige, or for denial of regional access against superior conventional forces. In addition, unless and until the current nuclear weapons states agree to disarm substantially, or entirely, their nuclear arsenals and weapons making complexes, the appeal of nuclear weapons as equalizers of the weak against the strong (in conventional forces) will be as powerful as is the contrarian, and paradoxical, appeal of nuclear weapons as tickets of admission to the ranks of great powers.

This paradox, of the simultaneous appeal of nuclear weapons to the (presumably) status quo oriented great powers, and to the challengers of existing political orders in various regions, makes the journey from the present condition toward the endgame of nuclear abolition a difficult one. The difficulty lies, not in the technical or scientific means for getting rid of nuclear weapons and their delivery systems, but in the politics of states' ambitions and fears. Thucydides argued that states fight for reasons of fear, honor and interest. Hobbes reminded us that, absent the existence of a Leviathan (or, in the case of international politics, a world government), a state of nature or legal anarchy exists in which survival depends on self help. Thus the shadow cast by nuclear weapons in the first decade of the twenty-first century is one of ambiguity. There is no imminent risk of global annihilation of the kind that was presented to policy makers during the High Cold War. Leaders nowadays fear climate change or other ecological disasters more than nuclear incineration. Additional reassurance can be found from the slow spread of nuclear weapons in the preceding century and, when all is said and done, in the present one. The existence of eight acknowledged and nine de facto nuclear weapons states is a more propitious outcome for international stability and peace than many Cold War pessimists had forecast, including some American Presidents and national security experts.

Precisely because politics is the dominant faculty and war the dependent variable, the preceding paragraph offers no brief for nuclear complacency. Nuclear weapons at rest, unless and until eliminated from existing arsenals, are potential instruments for peace and for war. They, like the atomic substructures from which they are built, exert both negative and positive forces on leaders of governments, on mass publics and on military planners. For example, consider giving the briefing on nuclear abolition for government and policy influential audiences in the national capitals of each of the five Permanent Members of the UN Security Council. After each session, the rediscovery of the "tragedy of the commons" would reappear. In a number of cases, if not all, nuclear altruism would be conditional on the batting order for disarmament: you first, then us.

This is not a cynical argument. Heads of state and their respective governments are charged, above all other responsibilities, with national security and defense. Moreover, and again, paradoxically from the standpoint of international peace and security, the "you first" position on arms control and disarmament can actually work in support of peace and stability. Although not uncontested, this is the argument of realist theorists of international politics: prudent self help, as opposed to reckless policies of imperialism or appeasement, can reduce the likelihood of war and the costs of war, should it occur. Since academics can and will argue about anything, the immediately preceding assertion lends itself to endless professional debate between theorists who are "offensive" versus "defensive" realists, but these nuances do not concern us here.

There is, on a more optimistic note, an answer to realists, with respect to the inevitability of national "you first" declarations with regard to nuclear disarmament or abolition. The answer is that the states most heavily endowed with nukes have a legal, moral and even strategic responsibility to start the process – to "engage" the issue and each other, in the fashionable diplomatic jargon of the present day. This means, first, that the United States and Russia have a special responsibility to lead the pack of nuclear weapons states in moving from the present condition to one that is less endowed with superfluous nuclear weapons, with nuclear weapons states, and with nuclear-aspiring candidates. Russian–American cooperation on nuclear arms reductions, on nonproliferation and (Cold Warriors, swallow hard) on missile defense is the necessary kick start for other nuclear weapons states to pile on. Conversely, a U.S. and Russian cop out from their leadership roles in arms reductions, nonproliferation and disarmament will stalemate the current condition of flux in the cosmos and eventually open the door to greater nuclear, and other, instability. Of course, stability has been oversold as a High Concept – it can be an excuse for lack of imagination and innovation in public policy, including arms control and military planning.

Research and development on technologies to make nuclear weapons "impotent and obsolete" in President Reagan's words will continue in the United States and elsewhere. Someday a convergence of space science, artificial intelligence, electromagnetic delivery systems, and nanotechnology may present the world with a powered-up nuclear neutralizer. Whether that world will be more peaceful than the present condition would be an interesting discussion, not only for academics, but also for governments and armed forces in nuclear weapons states and elsewhere. A world safe from nuclear war is not necessarily safe from war, and perhaps, more permissive of large scale conventional wars that some states, including Russia, now count on the nuclear weapons to deter. On the other hand, even short of such a technology "negation of the negation" for nuclear weapons, the unique awfulness of those weapons creates understandable temptations to urge forward their obsolescence. Whether we duck the option for nuclear abolition, skirt the edge of it, or actually accomplish it – the knowledge how to build nuclear weapons remains in our collective DNA, forever removing our innocence as to human ingenuity in the cause of self destruction. As Lawrence Freedman has noted: "There can be no purely nuclear strategies, but there remains a continuing need for strategies that take nuclear weapons into account."[3]

---

[3] Lawrence Freedman, *The Evolution of Nuclear Strategy*, Third Edition (New York: Palgrave Macmillan, 2003), p. 464.

Bibliography

Active Engagement, Modern Defence. "Strategic Concept for the Defence and Security of The Members of the North Atlantic Treaty Organization," Adopted by Heads of State and Government in Lisbon, http://www.nato.int/cps/en/natolive/official_texts_68580.htm.

Alberts, David S. *The Unintended Consequences of Information Age Technologies: Avoiding the Pitfalls, Seizing the Initiative.* Washington, DC: National Defense University, Institute for National Strategic Studies, Center for Advanced Concepts and Technology, April 1996.

Alberts, David S., John J. Garstka and Frederick P. Stein. *Network Centric Warfare: Developing and Leveraging Information Superiority.* Washington, DC: Command and Control Research Program, U.S. Department of Defense, 6th printing, April 2005.

Alberts, David S., John J. Garstka, Richard E. Hayes and David T. Signori. *Understanding Information Age Warfare.* Washington, DC: DOD Command and Control Research Program, U.S. Department of Defense, October 2004, Third Edition.

Albright, David and Jacqueline Shire with Paul Brannan and Andrea Scheel. *Nuclear Iran: Not Inevitable.* Washington, DC: Institute for Science and International Security, January 21, 2009.

Allison, Graham T. *Essence of Decision: Explaining the Cuban Missile Crisis.* Boston, MA: Little, Brown, 1971.

—. *Nuclear Terrorism: The Ultimate Preventable Catastrophe.* New York: Times Books – Henry Holt and Co., 2004.

Andrew, Christopher and Oleg Gordievsky. *KGB: The Inside Story of Its Foreign Operations from Lenin to Gorbachev.* New York: Harper Perennial, 1991.

Andrues, Wesley R. "What U.S. Cyber Command Must Do," *Joint Force Quarterly*, Issue 59 (4th Quarter 2010), 115–120.

Arbatov, Alexei. "Terms of Engagement: Weapons of Mass Destruction Proliferation and U.S.–Russian Relations," Ch. 5 in Stephen J. Blank, ed., *Prospects for U.S.–Russian Security Cooperation* (Carlisle, PA: U.S. Army War College, March 2009), pp. 139–168.

Arbatov, Alexei and Rose Gottemoeller. "New Presidents, New Agreements? Advancing U.S.–Russian Strategic Arms Control," *Arms Control Today*, July–August 2008, http://www.armscontrol.org/act/2008_07-08/CoverStory.asp.

Arquilla, John. *Worst Enemy: The Reluctant Transformation of the American Military.* Chicago, IL.: Ivan R. Dee, 2008.

Arquilla, John and David Ronfeldt. *Cyberwar Is Coming*! Santa Monica, CA: RAND, 1992.

— eds. *In Athena's Camp: Preparing for Conflict in the Information Age.* Santa Monica, CA: RAND, 1997.

Aslund, Anders and Andrew Kuchins. *The Russia Balance Sheet*. Washington, DC: Peterson Institute for International Economics and Center for Strategic and International Studies, April 2009.

Baker, Peter. "Russia and U.S. Sign Nuclear Arms Reduction Pact," *New York Times*, April 8, 2010, http://www.nytimes.com/2010/04/09/world/europe/09prexy.html.

Ball, Desmond. "The Development of the SIOP, 1960–1983," Ch. 3 in Desmond Ball and Jeffrey Richelson, eds., *Strategic Nuclear Targeting* (Ithaca, NY: Cornell University Press, 1986), pp. 57–83.

—. "U.S. Strategic Forces: How Would They Be Used?," in Steven E. Miller, ed., *Strategy and Nuclear Deterrence* (Princeton, NJ: Princeton University Press, 1984), pp. 215–244.

Barnett, Thomas P.M. *The Pentagon's New Map: War and Peace in the Twenty-first Century*. New York: Berkley Books, 2004.

Baylis, John and Robert O'Neill, eds. *Alternative Nuclear Futures: The Role of Nuclear Weapons in the Post-Cold War World*. Oxford: Oxford University Press, 2000.

Blair, Bruce G. *The Logic of Accidental Nuclear War*. Washington, DC: Brookings Institution, 1993.

Blair, Bruce, Victor Esin, Matthew McKinzie, Valery Yarynich and Pavel Zolotarev, "Smaller and Safer: A New Plan for Nuclear Postures," *Foreign Affairs*, No. 5 (September/October 2010), pp. 9–16.

Blank, Stephen J. *Russia and Arms Control: Are There Opportunities for the Obama Administration?* Carlisle, PA: Strategic Studies Institute, U.S. Army War College, March 2009.

— ed. *Prospects for U.S.–Russian Security Cooperation*. Carlisle, PA: U.S. Army War College, March 2009.

Blank, Stephen J. and Richard Weitz, eds. *The Russian Military Today and Tomorrow: Essays in Memory of Mary Fitzgerald*. Carlisle, PA: Strategic Studies Institute, U.S. Army War College, July 2010.

Blight, James G. and David A. Welch. *On the Brink: Americans and Soviets Reexamine the Cuban Missile Crisis*. New York: Hill and Wang, 1989.

Boot, Max. *War Made New: Technology, Warfare, and the Course of History, 1500 to Today*. New York: Gotham Books, 2006.

Bracken, Paul. *Fire in the East: The Rise of Asian Military Power and the Second Nuclear Age*. New York: Harper Collins, 1999.

Brodie, Bernard. *War and Politics*. New York: Macmillan, 1973.

Bundy, McGeorge, William J. Crowe, Jr. and Sidney D. Drell. *Reducing Nuclear Danger: The Road Away from the Brink*. New York: Council on Foreign Relations, 1993.

Center for Nonproliferation Research – National Defense University, and Center for Global Security Research, Lawrence Livermore National Laboratory. *U.S. Nuclear Policy in the 21st Century: A Fresh Look at National Strategy and Requirements*. Washington, DC: U.S. Government Printing Office, 1998.

Chalmers, Macolm and Simon Lunn. *NATO's Tactical Nuclear Dilemma*. London: Royal United Services Institute, Occasional Paper, March 2010, www.rusi.org.

Charap, Samuel and Mikhail Troitskiy, "Time to Put an End to MAD," *The Moscow Times*, January 18, 2011, http://www.themoscowtimes.com/opinion/article/time-to-put-an-end-to-mad/429055.html.

Cheng, Dean, "Chinese Views on Deterrence," *Joint Force Quarterly*, Issue 60 (1st Quarter 2011), pp. 92–94.

Choe Sang-Hun. "North Korea Says It Has 'Weaponized' Plutonium," *International Herald Tribune*, January 17, 2009, http://www.iht.com/articles/2009/01/17/news/norkor.1-409776.php.

—. "South Korea Publicly Blames the North for Ship's Sinking," *New York Times*, May 19, 2010, http://www.nytimes.com/2010/05/20/world/asia/20korea.html.

Cimbala, Stephen J. *Nuclear Weapons and Cooperative Security in the 21st Century.* London: Routledge, 2010.

Cimbala, Stephen J. and James Scouras. *A New Nuclear Century.* Westport, CT: Praeger Publishers, 2002.

Cirincione, Joseph. *Bomb Scare: The History and Future of Nuclear Weapons.* New York: Columbia University Press, 2007.

—. "Global Strikeout," *Foreign Policy*, April 23, 2010, http://www.foreignpolicy.com/articles/2010/04/23/global_strikeout.

Clarke, Richard A. and Robert K. Knake. *Cyber War.* New York: Harper Collins, 2010.

Collina, Tom Z. "Missile Defense Test a 'Success': Pentagon," *Arms Control Today*, May 2011, Arms Control Association, http://www.armscontrol.org/print/4860, downloaded May 6, 2011.

Cordesman, Anthony H. *The Korean Military Balance 2011: Comparative Korean Forces and the Forces of Key Neighboring States: Executive Summary.* Washington, DC: Center for Strategic and International Studies, revised May 6, 2011.

Daalder, Ivo and Jan Lodal. "The Logic of Zero," *Foreign Affairs*, No. 6 (November–December 2008), pp. 80–95.

de Haas, Marcel. "Russia's New Military Doctrine: A Compromise Document," in *Russian Analytical Digest*, No. 78, May 4, 2010, pp. 2–4, www.res.ethz.ch, downloaded December 15, 2010.

Dyson, Freeman. *Weapons and Hope.* New York: Harper Colophon Books, 1985.

Etzioni, Amitai. "Can a Nuclear-Armed Iran Be Deterred?," *Military Review*, May–June 2010, pp. 117–125.

Forsyth, James Wood, Jr., B. Chance Saltzman and Gary Schaub Jr. "Minimum Deterrence and Its Critics," *Strategic Studies Quarterly*, No. 4 (Winter, 2010), pp. 3–12.

—. "Remembrance of Things Past: The Enduring Value of Nuclear Weapons," *Strategic Studies Quarterly*, No. 1 (Spring, 2010), pp. 74–89.

Freedman, Lawrence. *The Evolution of Nuclear Strategy.* New York: St. Martin's Press, 1981 and 1983.

—. *The Evolution of Nuclear Strategy.* New York: Palgrave Macmillan, 2003, Third Edition.

—. *The Transformation of Strategic Affairs.* London: International Institute of Strategic Studies, Adelphi Paper 379, and Routledge Publishers, 2006.

Friedman, George. "The BMD Decision and the Global System," Stratfor.com, September 21, 2009, in *Johnson's Russia List* 2009 – #175, September 22, 2009, davidjohnson@starpower.net.

Gates, Robert M. "A Better Missile Defense for a Safer Europe," *New York Times*, September 19, 2009, http://www.nytimes.com/2009/09/20/opinion/20gates.html.

George, Alexander L. "A Provisional Theory of Crisis Management," in Alexander L. George, ed., *Avoiding War: Problems of Crisis Management* (Boulder, CO: Westview Press, 1991), pp. 22–27.

—. "The Tension Between 'Military Logic' and Requirements of Diplomacy in Crisis Management," in *Avoiding War: Problems of Crisis Management*, (Boulder, CO: Westview Press, 1991), pp. 13–21.

George, Alexander L. and William E. Simons, eds. *The Limits of Coercive Diplomacy*. Boulder, CO: Westview Press, 1994, Second Edition.

Giles, Keir. *The Military Doctrine of the Russian Federation 2010*, NATO Research Review. Rome: NATO Defense College, Research Division, February 2010.

Golts, Alexander, "Calling Moscow's Bluff on Missile Defense," *Moscow Times*, September 22, 2009, in *Johnson's Russia List* 2009 – #175, September 22, 2009, davidjohnson@starpower.net.

Goodman, Will. "Cyber Deterrence: Tougher in Theory than in Practice?," *Strategic Studies Quarterly*, No. 3 (Fall, 2010), pp. 102–135.

Gottemoeller, Rose. Assistant Secretary, Bureau of Verification, Compliance, and Implementation, U.S. Department of State, "The Long Road from Prague," Colonial Williamsburg, Virginia, August 14, 2009, http://www.state.gov/t/vci/rls/127958.htm.

Goure, Daniel, "Russian Strategic Nuclear Forces and Arms Control: Déjà vu All over Again," Ch. 5 in Stephen J. Blank and Richard Weitz, eds., *The Russian Military Today and Tomorrow: Essays in Memory of Mary Fitzgerald* (Carlisle, PA: Strategic Studies Institute, U.S. Army War College, July 2010), pp. 301–329.

Gray, Colin S. *Another Bloody Century: Future Warfare*. London: Weidenfeld and Nicolson, 2005.

—. "Geography and Grand Strategy," Ch. 9 in Colin S. Gray, *Strategy and History: Essays on Theory and Practice* (London: Routledge, 2006), pp. 137–150.

Gray, Colin S. *The Implications of Preemptive and Preventive War Doctrines: A Reconsideration*. Carlisle, PA: Strategic Studies Institute, U.S. Army War College, July 2007.

—. *The Second Nuclear Age*. Boulder, CO: Lynne Rienner, 1999.

—. *Strategy for Chaos: Revolutions in Military Affairs and the Evidence of History*. London: Frank Cass, 2002.

—. *War, Peace and International Relations: An Introduction to Strategic History*. New York: Routledge, 2007.

Hammes, Col. Thomas X., USMC (Ret.), "Information Warfare," Ch. 4 in G.J. David, Jr. and T.R. McKeldin III, eds., *Ideas as Weapons: Influence and Perception in Modern Warfare* (Washington, DC: Potomac Books, 2009), pp. 27–34.

Hecker, Siegfried S. *A Return Trip to North Korea's Yongbyon Complex*. Stanford, CA: Center for International Security and Cooperation, Stanford University, November 20, 2010.

—. "The Risks of North Korea's Nuclear Restart," *Bulletin of the Atomic Scientists*, May 12, 2009, http://thebulletin.org/web-edition/features/the-risks-of-north-koreas-nuclear-restart.

Hecker, Siegfried S. "What I Found in North Korea," *Foreign Affairs*, December 9, 2010, http://www.foreignaffairs.com/print/66970.

Henley, Lonnie D. "War Control: Chinese Concepts of Escalation Management," Ch. 5 in Andrew Scobell and Larry M. Wortzel, eds., *Shaping China's Security Environment: The Role of the People's Liberation Army* (Carlisle, PA: U.S. Army War College, Strategic Studies Institute, October 2006), pp. 81–103.

Herspring, Dale R. "Putin and Military Reform," Ch. 8 in Herspring, ed., *Putin's Russia: Past Imperfect, Future Uncertain* (Lanham, MD: Rowman and Littlefield, 2007), Third Edition, pp. 173–194.

— ed. *Putin's Russia: Past Imperfect, Future Uncertain*. Lanham, MD: Rowman and Littlefield, 2007, Third Edition.

—. "Russian Nuclear and Conventional Weapons: The Broken Relationship," paper presented at conference on "Strategy and Doctrine in Russian Security Policy," Ft. McNair, National Defense University, Washington, DC, June 28, 2010.

Hoffman, David E. *The Dead Hand: The Untold Story of the Cold War Arms Race and Its Dangerous Legacy*. New York: Doubleday, 2009.

Huntington, Samuel P. *The Clash of Civilizations and the Remaking of World Order*. New York: Touchstone Books, 1998.

Iran Watch. "Iran's Nuclear Timetable," updated December 2, 2010 and regularly, http://www.iranwatch.org/ourpubs/articles/iranucleartimetable.html.

Ivanov, Vladimir. "Comparison of Russian, U.S., and NATO Policies Toward Preemptive Nuclear Strikes," *Nezavisimoye Voyennoye Obozreniye*, February 4, 2008, in *Johnson's Russia List*, 2008 – #25, February 5, 2008, davidjohnson@starpower.net.

Jenkins, Brian Michael. *Will Terrorists Go Nuclear?* New York: Prometheus Books, 2008.

Jensen, Dr. Donald N. "Disagreements over Missile Defense Threaten to Undermine New START Treaty," *Voice of America Russian Service*, February 9, 2011, www.voanews.com/Russian/news in *Johnson's Russia List* 2011 – #25, February 11, 2011, davidjohnson@starpower.net.

Jervis, Robert. *The Meaning of the Nuclear Revolution: Statecraft and the Prospect of Armageddon*. Ithaca, NY: Cornell University Press, 1989.

Jing-dong Yuan. *Chinese Perceptions of Nuclear Weapons: Prospects and Potential Problems in Disarmament*. Proliferation Papers, No. 34. Paris: Security Studies Center, IFRI, Spring 2010, ifri@ifri.org.

Kahn, Herman. *On Escalation: Metaphors and Scenarios*. New York: Frederick A. Praeger, 1965.

—. *On Thermonuclear War*. New York: The Free Press, 1969, Second Edition.

Kamphausen, Roy and Andrew Scobell, eds. *Right-Sizing the People's Liberation Army: Exploring the Contours of China's Military*. Carlisle, PA: U.S. Army War College, Strategic Studies Institute, September 2007.

Karaganov, Sergei. "The START II Promotes Russian–American Cooperation and Rapprochement: Nuclear Free World Is a Dangerous Concept That Ought to be Abandoned," *Rossiiskaya Gazeta*, April 23, 2010, in *Johnson's Russia List* 2010 – #81, April 23, 2010, davidjohnson@starpower.net.

Kimball, Daryl G. "Next Steps on New START," *Arms Control Today*, April 2010, http://www.armscontrol.org, downloaded April 1, 2010.

Kipp, Jacob W. "Russia's Tactical Nuclear Weapons and Eurasian Security," *Jamestown Foundation Eurasia Defense Monitor*, March 5, 2010, in *Johnson's Russia List* 2010 – #46, March 8, 2010, davidjohnson@starpower.net.

Kislyakov, Andrei. "Russian Army Prepares for Nuclear Onslaught," *RIA Novosti*, January 29, 2008, in *Johnson's Russia List* 2008 – #20, January 29, 2008, davidjohnson@starpower.net.

Kissinger, Henry A. "Containing the Fire of the Gods," *International Herald Tribune*, February 6, 2009, http://www.iht.com/articles/2009/02/06/opinion/edkissinger.php.

Kramer, David J. "U.S. Abandoning Russia's Neighbors," *Washington Post*, May 15, 2010, in *Johnson's Russia List* 2010 – #96, May 17, 2010, davidjohnson@starpower.net.

Krepinevich, Andrew F. *7 Deadly Scenarios: A Military Futurist Explores War in the 21st Century*. New York: Bantam Books, 2009.

Krepon, Michael. *Better Safe than Sorry: The Ironies of Living with the Bomb*. Stanford, CA: Stanford University Press, 2009.

Kristensen, Hans M. *U.S. Nuclear Weapons in Europe: A Review of Post-Cold War Policy, Force Levels, and War Planning*. Washington, DC: Natural Resources Defense Council, February 2005.

Kuehl, Daniel T. "From Cyberspace to Cyberpower: Defining the Problem," Ch. 2 in Franklin D. Kramer, Stuart H. Starr and Larry K. Wentz, eds., *Cyberpower and National Security* (Washington, DC: National Defense University Press – Potomac Books, Inc., 2009), pp. 24–42.

Kuzmin, Vladimir. "From START to Euro-ABM," *Rossiyskaya Gazeta*, January 31, 2011, in *Johnson's Russia List* 2011 – #18, January 31–February 2, 2011, davidjohnson@starpower.net.

Lambeth, Belnjamin S. "Airpower, Spacepower, and Cyberpower," *Joint Force Quarterly*, Issue 60 (1st Quarter 2011), pp. 46–53.

Lebow, Richard Ned and Janice Gross Stein. *We All Lost the Cold War*. Princeton, NJ: Princeton University Press, 1994.

Levite, Ariel E. "Global Zero: An Israeli Vision of Realistic Idealism," *The Washington Quarterly*, April 2010, pp. 157–168, http://www.twq.com/10april/docs/10apr_Levite.pdf.

Lewis, George N. and Theodore A. Postol, "A Flawed and Dangerous U.S. Missile Defense Plan," *Arms Control Today*, May 2010, http://www.armscontrol.org/print/4244.

Libicki, Martin C. *Cyberdeterrence and Cyberwar*. Santa Monica, CA: RAND Corporation, 2009.

—. *Defending Cyberspace and other Metaphors*. Washington, DC: National Defense University, Directorate of Advanced Concepts, Technologies, and Information Strategies, February 1997.

Libicki, Martin C. *What Is Information Warfare?* Washington, DC: National Defense University, ACIS Paper 3, August 1995.

Lieber, Keir A. and Daryl G. Press, "The Nukes We Need: Preserving the American Deterrent," *Foreign Affairs*, No. 6 (November/December 2009), pp. 39–51.

MacFarquhar, Neil. "U.N. Approves New Sanctions to Deter Iran," *New York Times*, June 9, 2010, http://www.nytimes.com/2010/06/10/world/middleeast/10sanctions.html.

McDermott, Roger N. "Russia's Conventional Armed Forces, Reform and Nuclear Posture to 2020," Paper presented at conference on "Strategy and Doctrine in Russian Security Policy," Ft. McNair, National Defense University, Washington, DC, June 28, 2010.

McDonald, Mark. "Crisis Status in South Korea after North Shells Island," *New York Times*, November 23, 2010, http://www.nytimes.com/2010/11/24/world/asia/24korea.html.

—. "South Korea Digests News of North's Nuclear Site," *New York Times*, November 22, 2010, http://www.nytimes.com/2010/11/23/world/asia/23korea.html.

Medvedev, Dmitri, President of Russia, Address to the 64th Session of the UN General Assembly, New York, September 23, 2009, *Kremlin.ru*, in *Johnson's Russia List* 2009 – #177, September 24, 2009, davidjohnson@starpower.net.

"The Military Doctrine of the Russian Federation," text, www.Kremlin.ru, February 5, 2010, in *Johnson's Russia List* 2010 – #35, February 19, 2010, davidjohnson@starpower.net.

Miller, Franklin, George Robertson and Kori Schake. *Germany Opens Pandora's Box*. London: Centre for European Reform, February 2010, www.cer.org.uk.

Miller, Robert A., Daniel T. Kuehl and Irving Lachow, "Cyber War: Issues in Attack and Defense," *Joint Force Quarterly*, Issue 61 (2nd Quarter 2011), pp. 18–23.

Morgan, Patrick M. *Deterrence Now*. Cambridge: Cambridge University Press, 2003.

Mueller, John. "Think Again: Nuclear Weapons," *Foreign Policy*, January/February 2010, http://www.foreignpolicy.com/articles/2010/01/04/think_again_nuclear_weapons.

Mueller, Karl P., Jasen J. Castillo, Forrest E. Morgan, Negeen Pegahi, and Brian Rosen. *Striking First: Preemptive and Preventive Attack in U.S. National Security Policy*. Santa Monica, CA: RAND, 2006.

NATO. "Allied Leaders Agree on NATO Missile Defense System," November 20, 2010, http://www.nato.int/cps/en/SID-63044043-640CF22E/natolive/news_68439.htm.

—. *NATO 2020: Assured Security; Dynamic Engagement, Analysis and Recommendations of the Group of Experts on a New Strategic Concept for NATO*. Brussels: North Atlantic Treaty Organization, May 17, 2010.

—. "NATO Mission Statement Supports Retaining Tactical Nukes," *Global Security Newswire*, May 17, 2010, http://gsn.nti.org/gsn/nw_20100517_8029.php.

—. "NATO-Russia Set on Path Towards Strategic Partnership," November 20, 2010, http://www.nato.int/cps/en/natolive/news_68876.htm.

Norris, Robert S. "The Senate and the START Treaty," *Washington Times*, November 12, 2009, http://www.washingtontimes.com/news2009/nov/12/norris-the-senate-and-the-start-treaty/html.

Nunn, Sam. "NATO, Nuclear Security and the Terrorist Threat," *New York Times*, November 16, 2010, http://www.nytimes.com/2010/11/17/opinion/17/iht-ednunn.html.

Oliker, Olga, Keith Crane, Lowell H. Schwartz and Catherine Yusupov. *Russian Foreign Policy: Sources and Implications*. Santa Monica, CA: RAND Corporation, 2009.

Payne, Keith B. *Deterrence in the Second Nuclear Age*. Lexington, KY: University Press of Kentucky, 1996.

—. "Future of Deterrence: The Art of Defining How Much Is Enough," *Comparative Strategy*, No. 3 (July–August 2010), pp. 217–222.

Pellerin, Cheryl. "DARPA Goal for Cybersecurity: Change the Game," *American Forces Press Service*, December 20, 2010, http://www.af.mil/news/story_print.asp?id=123235799.

Perkovich, George. *Nuclear Weapons in Germany: Broaden and Deepen the Debate*. Washington, DC: Carnegie Endowment for International Peace, February 2010.

—. *The Obama Nuclear Agenda One Year after Prague*. Washington, DC: Carnegie Endowment for International Peace, March 31, 2010.

Pifer, Steven. "Beyond START: Negotiating the Next Step in U.S. and Russian Strategic Nuclear Arms Reductions," Brookings Institution, www.brookings.edu, in *Johnson's Russia List* 2009 – #88, May 12, 2009, davidjohnson@starpower.net.

Pikayev, Alexander L. "For the Benefit of All," *Moscow Times*, September 21, 2009, in *Johnson's Russia List* 2009 – #174, September 21, 2009, davidjohnson@starpower.net.

Podvig, Pavel. *Blog on Russian Strategic Nuclear Forces* (ongoing), http://russianforces.org/.

—. "New START Treaty in Numbers," from his blog, *Russian Strategic Nuclear Forces*, April 9, 2010, http://russianforces.org/blog/2010/03/new_start_treaty_in_numbers.shtml.

—. "Russia's New Arms Development," *Bulletin of the Atomic Scientists*, January 16, 2009, http://thebulletin.org/web-edition/columnists/pavel-podvig/russias-new-arms-development.

—. "What If North Korea Were the Only Nuclear Weapon State?," *Bulletin of the Atomic Scientists*, May 27, 2009, http://thebulletin.org/web-edition/columnists/pavel-podvig/.

—. "What To Do About Tactical Nuclear Weapons," *Bulletin of the Atomic Scientists*, February 25, 2010, http://the bulletin.org, in *Johnson's Russia List* 2010 – #43, March 3, 2010, davidjohnson@starpower.net.

Pollack, Jonathan D. *North Korea's Nuclear Weapons Development: Implications for Future Policy*. Proliferation Papers, No. 33. Paris: Security Studies Center, IFRI, Spring 2010, esp. p. 35, http://ifri.org/downloads/pp33pollack.pdf.

Pollpeter, Kevin. "Towards an Integrative C4ISR System: Informationization and Joint Operations in the People's Liberation Army," Ch. 5 in Roy Kamphausen, David Lai and Andrew Scobell, eds., *The PLA at Home and Abroad: Assessing the Operational Capabilities of China's Military* (Carlisle, PA: Strategic Studies Institute, U.S. Army War College, June 2010), pp. 193–235.

Quester, George H. *Nuclear First Strike: Consequences of a Broken Taboo*. Baltimore, MD: Johns Hopkins University Press, 2006.

Rasmussen, Anders Fogh. "NATO's Common European Roof," *Moscow Times*, April 22, 2010, in *Johnson's Russia List* 2010 – #83, April 29, 2010, davidjohnson@starpower.net.

Rothman, Alexander H. and Lawrence J. Korb. "Pakistan Doubles Its Nuclear Arsenal: Is It Time to Start Worrying?," *Bulletin of the Atomic Scientists*,

February 11, 2011, http://www.thebulletin.org/node/8607, downloaded February 15, 2011.

"Russia's Message on Reshaping Its Nuclear Doctrine," Stratfor.com, October 15, 2009, in *Johnson's Russia List* 2009 – #190, October 15, 2009, davidjohnson@starpower.net.

Sanger, David E. "After Treaty, Obama Nuclear Agenda Only Gets Harder," *New York Times*, December 21, 2010, http://www.nytimes.com/2010/12/22/us/politics/22assess.html.

Sanger, David E. and Joseph Berger, "Arms Bid Seen in New N. Korea Plant," *New York Times*, November 21, 2010, http://www.nytimes.com/2010/11/22/us/22talk.html.

Sanger, David E. and Choe Sang-Hun. "North Korea Cuts All Ties with South," *New York Times*, May 25, 2010, http://www.nytimes.com/2010/05/26/world/asia/26korea.html.

SarDesai, D.R. and Raju G.C. Thomas, eds. *Nuclear India in the Twenty-First Century.* New York: Palgrave Macmillan, 2002.

Schell, Jonathan. *The Seventh Decade: The New Shape of Nuclear Danger.* New York: Henry Holt and Co., 2007.

Schelling, Thomas C. *Arms and Influence.* New Haven, CT: Yale University Press, 1966.

—. "A World Without Nuclear Weapons?," *Daedalus*, No. 4 (Fall, 2009), 124–129.

Schulte, Paul. *Is NATO's Nuclear Deterrence Policy a Relic of the Cold War?* Washington, DC: Carnegie Endowment for International Peace, Policy Outlook, November 17, 2010.

Scobell, Andrew. "Discourse in 3-D: The PLA's Evolving Doctrine, Circa 2009," Ch. 3 in Roy Kampenhausen, David Lai and Andrew Scobell, eds., *The PLA at Home and Abroad: Assessing the Operational Capabilities of China's Military* (Carlisle, PA: Strategic Studies Institute, U.S. Army War College, June 2010), pp. 99–133.

—. *Projecting Pyongyang: The Future of North Korea's Kim Jong Il Regime.* Carlisle, PA: Strategic Studies Institute, U.S. Army War College, March 2008.

Scobell, Andrew and John M. Sanford. *North Korea's Military Threat: Pyongyang's Conventional Forces, Weapons of Mass Destruction, and Ballistic Missiles.* Carlisle, PA: U.S. Army War College, Strategic Studies Institute, April 2007.

Scowcroft, Brent, Joseph Nye, Nicholas Burns and Strobe Talbott. "U.S., Russia Must Lead on Arms Control," Politico.com, October 13, 2009, http://dyn.politico.com/printstory.cfm?

Sessler, Andrew M., John M. Cornwall, Bob Dietz, Steve Fetter, Sherman Frankel, Richard L. Garwin, Kurt Gottfried, Lisbeth Gronlund, George N. Lewis, Theodore A. Postol, and David C. Wright. *Countermeasures: A Technical Evaluation of the Operational Effectiveness of the Planned US National Missile Defense System.* Cambridge, MA: Union of Concerned Scientists, April 2000.

Shear, Michael D. and Ann Scott Tyson. "Obama Shifts Focus of Missile Shield," *Washington Post*, September 18, 2009, http://www.washingtonpost.com/wp-dyn/content/article/2009/09/17/AR2009091700639.html.

Sherr, James. "NATO and Russia: 'Refresh' but No Transformation," *Chatham House*, www.chathamhouse.org.uk, November 10, 2010, in *Johnson's Russia List* 2010 – #218, November 22, 2010.

Shultz, George P., William J. Perry, Henry A. Kissinger and Sam Nunn. "Toward a Nuclear-Free World," *Wall Street Journal*, January 15, 2008, A13.

Sigal, Leon V. "Let's Make a Deal," *The American Interest Magazine*, January–February 2010, http://the-american-interest.com/article-bd.cfm?piece=767.

Sokolski, Henry D., ed. *Pakistan's Nuclear Future: Worries Beyond War*. Carlisle, PA: Strategic Studies Institute, U.S. Army War College, January 2008.

— ed. *Reviewing the Nuclear Nonproliferation Treaty (NPT)*. Carlisle, PA: Strategic Studies Institute, U.S. Army War College, May 2010.

Sokov, Nikolai. "The New, 2010 Russian Military Doctrine: The Nuclear Angle," Center for Nonproliferation Studies, Monterey Institute of International Studies, February 5, 2010, http://cns.miis.edu/stories/100205_russian_nuclear_doctrine.htm.

—. "Nuclear Weapons in Russian National Security Strategy," Paper presented at conference on "Strategy and Doctrine in Russian Security Policy," Ft. McNair, National Defense University, Washington, DC, June 28, 2010.

"S. Korea, U.S. Launch Joint Committee to Deter N. Korea's Nuclear Threats," *Yonhap News Agency*, Seoul, December 13, 2010, http://app.yonhapnews.co.kr/YNA/Basic/Article/Print?YIBW_showEnArticlePrintView, downloaded December 15, 2010.

STRATFOR. "Russia's Message on Reshaping Its Nuclear Doctrine," Stratfor.com, October 15, 2009, in *Johnson's Russia List* 2009 – #190, October 15, 2009, davidjohnson@starpower.net.

Talbott, Strobe. "A Better Base for Cutting Nuclear Weapons," *Financial Times*, September 21, 2009, in *Johnson's Russia List* 2009 – #174, September 21, 2009, davidjohnson@starpower.net.

Tannenwald, Nina. *The Nuclear Taboo: The United States and the Non-Use of Nuclear Weapons Since 1945*. Cambridge: Cambridge University Press, 2007.

The White House. *National Security Strategy*. Washington, DC: The White House, May, 2010.

Thomas, Timothy L. "Russia's Asymmetrical Approach to Information Warfare," Ch. 5 in Stephen J. Cimbala, ed., *The Russian Military into the Twenty-first Century* (London: Frank Cass, 2001), pp. 97–121.

—. "Russian Information Warfare Theory: The Consequences of August 2008," Ch. 4 in Stephen J. Blank and Richard Weitz, eds., *The Russian Military Today and Tomorrow: Essays in Memory of Mary Fitzgerald* (Carlisle, PA: Strategic Studies Institute, U.S. Army War College, July 2010), pp. 265–299.

Tom Z. Collina, Tom Z. with Daryl G. Kimball. *Now More than Ever: The Case for the Comprehensive Nuclear Test Ban Treaty*. Washington, DC: Arms Control Association, Briefing Book, February 2010.

Toukan, Abdullah and Anthony Cordesman. *Iran, Israel and the Effects of a Nuclear Conflict in the Middle East*. Washington, DC: Center for Strategic and International Studies, June 1, 2009.

Traynor, Ian. "Pre-emptive Nuclear Strike a Key Option, NATO Told," *Guardian Unlimited*, January 22, 2008, http://www.guardian.co.uk/nato/story/0,,2244782,00.html.

Treaty Between the United States of America and the Russian Federation on Measures for the Further Reduction and Limitation of Strategic Offensive Arms (Washington, DC: U.S. Department of State, April 8, 2010), http://www.state.gov/documents/organization/140035.pdf.

Trenin, Dmitri. "Turning a Happy Hour into a Happy Alliance," *Moscow Times*, November 22, 2010, in *Johnson's Russia List* 2010 – #218, November 22, 2010.

Unclassified Statement of Lieutenant General Patrick J. O'Reilly, USA, Director, Missile Defense Agency, Before the House Armed Services Committee. Washington, DC: U.S. House of Representatives, House Armed Services Committee, October 1, 2009.

Unclassified Statement of Lt. Gen. Patrick J. O'Reilly, Director, Missile Defense Agency, Before the House Armed Services Committee, Subcommittee on Strategic Forces, Regarding the *Fiscal Year 2011 Missile Defense Programs*. Washington, DC: House Armed Services Committee, U.S. House of Representatives, April 15, 2010.

Wagner, Daniel and Diana Stellman. "The Prospects for Missile Defense Cooperation Between NATO and Russia," *Foreign Policy Journal*, February 10, 2011, www.foreignpolicyjournal.com in *Johnson's Russia l ist* 2011 – #24, February 10, 2011, davidjohnson@starpower.net.

Walt, Stephen M. "All the Nukes You Can Use," foreignpolicy.com, May 24, 2010, http://walt.foreignpolicy.com/posts/2010/05/24/all_the_nukes_that_you_can_use.

Waltz, Kenneth N. "More May Be Better," Ch. 1 in Scott D. Sagan and Kenneth N. Waltz, eds., *The Spread of Nuclear Weapons: A Debate* (New York: W.W. Norton, 1995), pp. 1–45, esp. p. 42.

Weitz, Richard. "Strategy and Doctrine in Russian Security Policy," Paper presented at conference on "Strategy and Doctrine in Russian Security Policy," Ft. McNair, National Defense University, Washington, DC, June 28, 2010.

Whitlock, Craig. "U.S. Developing New Non-nuclear Missiles," *Washington Post*, April 8, 2010, http://www.msnbc.msn.com/id/36253190/ns/us_news-washington_post/print/1/.

Wilkening, Dean. *Ballistic-Missile Defence and Strategic Stability*. Oxford: Oxford University Press, 2000.

Wortzel, Larry M. *China's Nuclear Forces: Operations, Training, Doctrine, Command, Control, and Campaign Planning*. Carlisle, PA: U.S. Army War College, Strategic Studies Institute, May 2007.

Young Whan Kihl and Hong Nack Kim, eds. *North Korea: The Politics of Regime Survival*. Armonk, NY: M.E. Sharpe, 2006.

Index